CHARACTERIZATION
OF
NANOSTRUCTURES

CHARACTERIZATION
OF
NANOSTRUCTURES

SVERRE MYHRA • JOHN C. RIVIÈRE

CRC Press
Taylor & Francis Group
Boca Raton   London   New York

CRC Press is an imprint of the
Taylor & Francis Group, an **informa** business

CRC Press
Taylor & Francis Group
6000 Broken Sound Parkway NW, Suite 300
Boca Raton, FL 33487-2742

First issued in paperback 2016

Version Date: 20120227

ISBN 13: 978-1-138-19863-0 (pbk)
ISBN 13: 978-1-4398-5415-0 (hbk)

Library of Congress Cataloging-in-Publication Data

Myhra, S. (Sverre), 1943-
   Charaterization of nanostructures / Sverre Myhra and John C. Riviere.
      p. cm.
   "A CRC title."
   Includes bibliographical references and index.
   ISBN 978-1-4398-5415-0 (alk. paper)
   1. Nanostructured materials. 2. Nanostructures--Optical properties. 3.
Nanotechnology. 4. Imaging systems. I. Rivière, J. C. II. Title.

TA418.9.N35M94 2012
620.1'15--dc23   2012001213

Visit the Taylor & Francis Web site at
http://www.taylorandfrancis.com

and the CRC Press Web site at
http://www.crcpress.com

Contents

SECTION I Techniques and Methods

SECTION II   Applications

Preface

A cursory search of the Web sites of the main publishers of scientific and technical literature reveals that some 55+ monographs with 'nano' in the title have been published in recent years. No doubt many more have been commissioned and are being written. When we first contemplated the idea of adding to the nanoliterature, it was by no means obvious that the field needed yet another monograph. So we deployed a web-based sociometric method, to find out the extent to which nanotechnology (we use the term as an ill-defined umbrella that covers all things with a nanoassociation), is popular and/or important, and how it rates in comparison with other important indicators of cultural or scientific significance. Our colleagues in the social sciences are beginning to make use of the Google™ search engine as a tool for tapping into, and mining, the greatest database of relationships, attitudes, and values that has ever been assembled—the Internet. The Internet is now the closest thing we have to a universal collective consciousness. The objective of our search was to decide whether, or not, our proposed monograph would make a worthwhile contribution.

The results (based on an unscientific search) are listed in Table 1.

Some interesting observations can be made. We have chosen 'quantum mechanics' and 'Shakespeare' as the respective sociometric control indicators for the impact of a well-established body of scientific knowledge and of the much broader influence of great works of art on the near-universal cultural consciousness. It is comforting that 'Shakespeare' outranks 'nanotechnology' by a factor of five. We also think that it is a healthy sign that 'nano hype' runs neck-and-neck with 'nanoscience'.

The ratio of hits for the key words (1) to the hits that include the addition of 'book' (2) was of particular interest, when we were thinking about committing ourselves to the task ahead. While the respective ratios for 'nanotechnology' and 'Shakespeare' are similar (ca. 3:1), it seems that the monograph literature for 'quantum mechanics' is oversubscribed by roughly a factor of ten (ca. 1:3). On the other hand, 'nanocharacterization' would appear to be lagging the field (ca. 500:1). We take this to mean that we are not wasting our time and energy. On the other hand, it could mean that those members of the nanocommunity who engage in materials characterization are the only ones who know what they are doing, but are too busy to write books.

We, the two authors, have recently celebrated a century of combined involvement in the broad field of materials characterization. It is only in recent times that we have come to appreciate that many of the problems that preoccupied us several decades ago would now be considered to merit the attachment of the nanobadge. During that time, our work has coincided with the birth and subsequent maturing of some of the most important characterization tools, such as surface and interface analysis in the 1960s, Raman spectroscopy in the '70s and '80s, and scanning probe microscopies in later years. We have also had the good fortune to have worked in some exceptional environments, ranging from the Harwell Laboratory of AEA Technology to the Department of Materials at the University of Oxford; these were places where some of the pioneering work on the electron-optical and spectroscopic techniques was

TABLE 1

The 'Hits' Delivered by a Google Search

	1 Key Word(s) Hits ($\times 10^6$)	2 Key Word(s) + Book Hits ($\times 10^6$)	Ratio 1/2
Nano*	261	4	65
Nanotechnology	23	9.2	2.5
Nanoscience	3	0.7	4
Nano Characterization	6	0.012	500
Nano Hype	2	1.4	1.5
Quantum Mechanics	3	10	0.3
Shakespeare	124	60	2

* Here the number of hits is flattered by the attachment of the term 'nano' to mass-consumption products, such as a well-known electronic gadget and a very small car, that have, at best, a tenuous connection with nano-technology. The search term nanotechnology is likely to be a more representative choice.

carried out. The breadth and depth of facilities and expertise that have surrounded us during our working lives rate among the very best there are. Being surrounded by industrious and gifted colleagues has given us a broader and richer perspective on materials characterization at all size scales, if for no other reason than by a process of osmosis.

THE STRUCTURE OF THE MONOGRAPH

The monograph commences with a brief introduction to the topic of materials characterization, in which the aims, the perceived needs of the readership, and how those needs can be met, are discussed. As well, there is some advice on how to make best use of the material in the monograph. The remainder of the volume can be divided into two parts, the first providing thumb-nail sketches of the most widely used techniques and methods that apply to nanostructures, and the second describing typical applications to single nanoscale objects, as well as ensembles of such objects. A concluding chapter provides an overview of the strategies for interrogating nano-structures, and mentions some of the tactical issues such as specimen preparation.

SECTION I: TECHNIQUES AND METHODS

In the six chapters in Section I, an overview is given of the physical principles of the main techniques, and a description of those operational modes that are most relevant to nanoscale characterization. Sufficient technical detail is given to enable the reader, and prospective user, to gain an appreciation for the strengths and limitations of particular techniques. The chapters deal with the mainstream techniques, as well as some used less commonly.

SECTION II: APPLICATIONS OF TECHNIQUES TO STRUCTURES OF DIFFERENT DIMENSIONALITIES AND FUNCTIONALITIES

The five chapters in Section II deal with methods for materials characterization of generic types of systems, using carefully chosen illustrations from the literature. A brief description of the material(s) is included in the introduction to each chapter in order to provide a context for the methods for characterization. Particular emphasis is placed on synthesis routes since our assessment is that that is where the current centre of gravity in the nanofield lies. Thus, materials characterization is of critical importance for the evaluation of products during synthesis, and for the validation of the resultant end-product, whether or not it was the one intended.

Due to the volume of the literature under the broad heading of nanotechnology, the treatment cannot be encyclopaedic, but will attempt to give justice to those accounts that have stood the test of time for systems of greatest current relevance. The focus will be on structures that are nanoscale in one, two, or three dimensions, and there will be particular emphasis on those structures that are carbon-based.

THE INTENDED AUDIENCE

The monograph is not intended to add to the knowledge of specialists in their areas of speciality. Rather it is intended to bridge the gap between the generalists, who play vital roles in the postdisciplinary area of nanotechnology, and the specialists, who view themselves more in the context of a discipline. It can also serve to facilitate communication between specialists who wish to meet on neutral ground. It could constitute useful reading for new entrants to the field, or as source material for a course on nanotechnology.

Acknowledgements

The list of colleagues and students who have contributed to our enjoyment and fascination with science, and who have added to our understanding and insights into materials characterization could occupy a slim volume in its own right. However, we wish to record our appreciation to colleagues attached to the Oxford Materials Characterization Service and BegbrokeNano, Dr. Alison Crossley, Dr. Colin Johnston, and Dr. Kerstin Jurkschat. Other colleagues and friends have read some of the chapters, and made valuable suggestions; they are acknowledged separately at the ends of the relevant chapters.

We also very much appreciate the institutional support offered by the head of the department of materials at the University of Oxford, Professor Chris Grovenor, and by the academic head of the Oxford University Begbroke Science Park, Professor Peter Dobson.

It is traditional to admit to human frailty, and we wish to uphold that tradition. We do not wish to claim completeness of coverage of the diverse and evolving field of nanotechnology. The content is due entirely to the idiosyncratic choices of the authors, and should not be construed to be a value-judgment of any other equally excellent, but absent in the monograph, work in the field. Likewise, we are prepared to accept total ownership and complete responsibility for any errors that might have crept in. We would of course be delighted to receive any communication pointing out any such errors.

Inevitably, the task of writing a monograph will entail devoting time and energy that could have been deployed in other ways. Our families and friends will know what we mean, and we thank them for their understanding of, and patience with, the foibles of two ageing scientists.

Sverre Myhra
John Rivière

The Authors

S. (Sverre) Myhra is currently a visiting scientist in the Department of Materials at the University of Oxford. Previously, he founded and headed an early scanning probe microscopy group in Australia. He is the author of numerous scholarly papers and chapters in the area of materials science and technology focusing on surface and interface analysis, with a recent emphasis on applications of scanning probe microscopy. He received the PhD (1968) in physics from the University of Utah, Salt Lake City.

J. (John) C. Rivière worked for AEA Technology, Harwell, England until his retirement. He is the author or coauthor of numerous scholarly papers and chapters that reflect his research interests in both basic and applied surface science and technology. A pioneer in the early development of electron spectroscopies, he was the recipient of the UK Vacuum Council Medal (1989). He received the MSc degree (1950) from the University of Western Australia, and the PhD (1955) and DSc (1995) degrees from the University of Bristol.

1 Introduction to Characterization of Nanostructures

1.1 NANOTECHNOLOGY—IN THE BEGINNING THERE WAS THE IDEA

The idea of nanotechnology is ascribed traditionally to the after-dinner speech given by Richard Feynman at the American Physical Society meeting on December 29th in 1959, entitled 'There is plenty of room at the bottom'. The excerpts below illustrate two critical requirements, italicised, identified by Feynman for turning his vision into reality.

> ... The problems of manufacture and reproduction will be quite different. I am, as I said, inspired by the biological phenomena in which chemical forces are used in a repetitious fashion to produce all kinds of weird effects (one of which is the author).... The principles of physics, as far as I can see, do not speak against *the possibility of manoeuvring things atom by atom*. It is not an attempt to violate any laws; it is something, in principle, that can be done.... The problems of chemistry and biology can be greatly helped if our *ability to see what we are doing*, and to do things on the atomic level, is ultimately developed—a development which I think cannot be avoided...

On the one hand, he asked for the tools to enable matter to be manipulated on the atomic scale, presumably with a spatial precision commensurate with the range of interatomic forces, while on the other, he also asked for the 'ability to see what we are doing'. It is in fact this latter requirement that constitutes the most important and relevant justification for this monograph. The requirement can be broadened, and at the same time be made more specific, by asking for the ability to interrogate physical systems at a level where the information volume is comparable to that of a single atom.

1.2 NANOTECHNOLOGY AS A PRACTICAL PROPOSITION

While the ideas of nanotechnology, as well as the critical requirements for its implementation, were formulated some 50 years ago, it took another 30 years before practicality caught up with the idea. The schematic below shows, in broad terms, how advances in our ability to characterize materials in 'real' experiments have gone hand-in-hand with the gradual emergence of nanotechnology as a practical proposition. Indeed, one could infer from the schematic that those advances were necessary precursors. The schematic also shows that a parallel trend, that of the ability to carry out numerically intensive *ab initio* computation on ever greater ensembles of atoms,

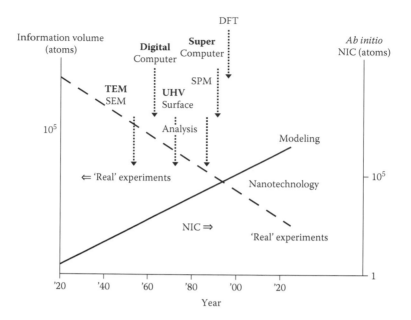

FIGURE 1.1 Schematic illustration of trends in numerically intensive computation (NIC), (as defined by the number of interacting atoms that can be subjected to *ab initio* numerical calculation, ascending solid line), and of the ability to do 'real' experiments on increasingly small information volumes (where the limit for 'real' experiments arises from the spatial resolution of the interrogating probe, descending broken line). The two trends are correlated with the emergence of theoretical and experimental tools (DFT = Density Functional Theory). The cross-over of the two trends, as defined by the approximate point of intersection at which the information volumes of analytical techniques became comparable to the size of an assembly of atoms that could be modelled, has defined the arrival of nanotechnology. (However one chooses to define nanotechnology, it emerged as a potentially practical proposition during the 1990s.)

has allowed theory to deal with systems that can be interrogated in 'real' experiments relevant to nanotechnology (see Figure 1.1).

1.3   WHAT IS NANOTECHNOLOGY?

Activities undertaken under the heading of 'technology' are generally thought to be concerned with an intention to arrive at an end product of some industrial commercial value. The activities may build on an empirical basis, as in the case of mediaeval coloured glass, where trial and error were found to produce effects that were subsequently understood to be associated with nanoparticles (Walters and Parkin 2009; Stained Glass Association of America 2010). In recent times, it has become the norm that a particular technology must be based on sound science. Thus many, maybe most, activities that are currently said to be nanotechnology should more properly be thought of as nanoscience. This is not to detract from the potential importance of some of the best research being carried out, whatever its label.

There are at least two prevailing views about the definition of nanotechnology. One view focuses mainly on the size scale, not only on the properties of matter that are emergent by virtue of the smallness of the size of an object, but also on the likely technical advances required to exploit these properties. In particular, this view and approach has it that if an object is taken with a particular functionality, and made smaller, while its functionality is retained, then that would lead automatically to a technical advance. The prime example is that of the gradual evolution from the macro to the nanoscale of electronic devices (see below). There are many other examples, including sensors and composites.

The other view focuses on the emergent richness of complexity when relatively simple, and often identical, nanoscale objects are combined into objects of arbitrary size. In this view, it is the process that is the focus, whereby an exceedingly large number of nanoscale units can be assembled by a design in such a manner that any degree of complexity can be achieved. The digital computer is an obvious example. However, nature provides any number of examples ranging from that of the present state of the universe being assembled from a relatively small number of types of fundamental particles, to that of familiar objects being constructed from the hundred or so types of atoms in the periodic table.

The two views are known as 'top-down' and 'bottom-up'. It is useful to explore the definitions in more detail (more exhaustive explorations can be found in the literature, see, e.g., Drexler 1992, for an early account).

1.3.1 NANOTECHNOLOGY—TOP-DOWN

The general thrust of the top-down definition is that nanotechnology is concerned principally with new phenomena encountered on the nanoscale, giving rise to novel applications and to new classes of devices. A criticism of this definition is that one, or even a small number, of such devices cannot have any significant industrial impact, unless the production process operates in parallel on a massive scale. (Parallel methods are the basis of all mass-produced consumer products. For example, the annual output of ca. 10^8 cars rolls off ca. 10^3 conveyor belts around the world. On the other hand, the global annual production of TiO_2 nanoparticles, performed by liquid phase synthesis, requires an enormous increase in parallel production.) And, the argument goes, complexity is difficult to engineer on the nanoscale. Thus, a useful nanotechnology can arise only if complexity can be an emergent attribute of the purposeful interaction between large numbers of nanoscale objects.

A prime example of top-down nanotechnology in action is that of the semiconductor device industry. The trend from micro to nano is illustrated in Figure 1.2. While the physical size of a central processing unit (CPU) (the heart of a personal computer) chip has remained substantially constant, the complexity in terms of transistors per chip has increased by a factor of 10^6 in the period 1971–2008. The basic building block, the NAND gate (a basic decision-making element in digital electronics), has decreased in size into the nano regime, and is approaching 20 nm. The cost of a CPU chip has remained substantially constant, due to a vast increase in the scale of parallel production, as well as to economy of scale.

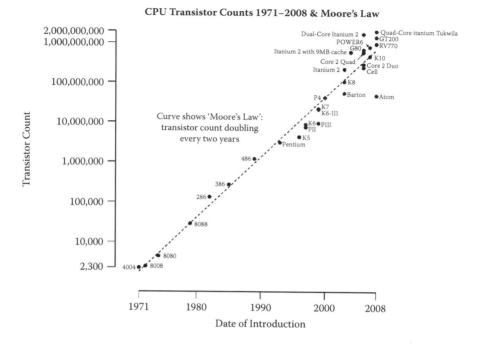

FIGURE 1.2 Evolution in complexity of a CPU chip, illustrating the transition from micro- to nanoscale devices by a top-down route. The trend, dotted line, shows that the transistor count has been doubling every two years, and was predicted by, and became known as *Moore's Law*. (This is an updated version based upon Moore (1965).)

While the trend from the micro- to the nanoscale illustrates the top-down process, it also demonstrates that the bottom-up process of constructing a large object, a CPU, from many nanoscale objects, that is, NAND gates, generates useful complexity by the carefully designed assembly of basic building blocks (see Figure 1.3). This leads to the alternative definition of nanotechnology (see the following section).

In parallel with the refinements that have paved the way for nanoscale process technologies, the industry has been a major driver, and user, of developments in instrumentation and methods for nanoscale characterization.

Progress as predicted by Moore's Law is unlikely to continue unless new phenomena are investigated and harnessed by industry (e.g., molecular electronics). In this sense, device technology will remain top-down.

1.3.2 NANOTECHNOLOGY—BOTTOM-UP

The implicit assumption here is that nanoscale objects are not *per se* particularly useful, although there is a great deal of current interest in establishing their properties, but that the development of process routes for production of bulk quantities is the critical issue, followed by methods for incorporating the objects and their properties into structures of arbitrary size. As bulk commodities, nanoscale objects are no more interesting, or valuable, than bulk quantities of TiO_2 particles for the paint industry.

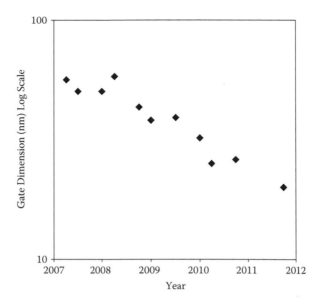

FIGURE 1.3 Recent trend in the physical size of the basic building block of logic circuits, the NAND gate, illustrating that the computer industry has arrived in the nano regime by a top-down process. The data show the dates of introduction of products by several major manufacturers. The data point at ca. 20 nm is a forecast.

Only when the basic building blocks are assembled from the bottom-up according to one or other intelligent design, do they become useful and valuable, and become capable of becoming the drivers of another generation of 'disruptive' technologies, similar to that of semiconductor logic.

The term 'disruptive technology' is used widely to describe a technology that helps to create a new market and value network, but eventually disrupts an existing market and value network, thereby displacing an earlier technology; the use of the term was popularized by Harvard Business School Professor C. Christensen in his book *The Innovator's Dilemma* (Christensen 1997). There are many recent examples of the effects of a disruptive technology, for example, the ubiquitous deployment of information technology. For instance, when the PC became part of the furniture of virtually all offices, then the profession of typist essentially ceased to exist (and even ageing scientists had to acquire rudimentary keyboard skills!). Likewise, the pre-IT business models of the sellers of books and music failed once online commerce and free (pirated?) downloads took off. The first effect of this was that many independent retailers were forced out of the market, while the larger chains are currently succumbing to the disruption of having to reinvent themselves or perish.

As stated above, the world, indeed the universe, as we know it, is the product of bottom-up technology, beginning with the Big Bang, and then, according to natural laws, evolving from the most basic building blocks through differentiation into a diverse and complex system. Only in the last 100 years have the theoretical and experimental tools for the interrogation, description, and explanation of the structure of the universe been acquired.

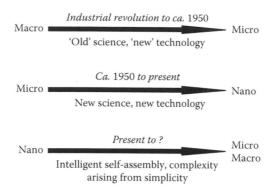

FIGURE 1.4 Progress of technology from macro to micro to nano, from the industrial revolution to the present day.

The emergence of life a few billion years ago brought another form of bottom-up technology into being, that is, natural selection, which became the driving force for differentiation and evolution of complexity. On the one hand, natural selection has been profoundly disruptive for all those species that did not adapt and therefore fell by the wayside, becoming at best part of the fossil record. On the other hand, some species have been successful in adapting to changing circumstances, largely through lucky dips in the pool of mutations, leading to survival by adoption of clever nanotechnologies. Thus, there is a largely untapped treasure trove, free for the taking, of natural nanotechnologies. These technologies have proved their utility, as judged by natural selection and fitness for purpose. The tools necessary for their characterization and categorization, and for the assembly of an inventory, have only just been realised. Nevertheless, a new field has been born, known as biomimetics (the learning from, and mimicking of, natural technologies).

While nanoscale technology at present is top-down and is dominated by the semiconductor industry, future technologies are likely to be of the bottom-up variety, as shown in Figure 1.4.

A single nanostructure must by its very nature be simple, and thus of little interest in itself (e.g., a single NAND gate). However, when a very large number of NAND gates is assembled into a computer, then complexity and utility become emergent qualities, illustrating the ultimate aim of the coming nanotechnological revolution. The post-war years have set the stage for a top-down transition from micro to nano. The future will witness a general disruptive transition from nanoscale to an arbitrary size-scale by bottom-up process technologies.

1.3.3 Nanotechnology—A Socio-Economic Definition

It is possible to define an activity in terms of its participants, and of their ethos and group allegiances. For instance, the participants in scientific activities will, as a group, define science in a way that sets those activities apart from those that define, for example, engineering. In Table 1.1, an attempt has been made to categorize how one might think about the various activities that are relevant to the idea of nanotechnology.

TABLE 1.1

Science and Technology: Theory, Practice, Setting, and Outcome

	Theory	Practice	Setting	Outcome
Science	Bottom-up	Top-down	Discipline	Explanation
Technology	Top-down	Top-down	Profession	Product process
Nanoscience	Bottom-up	Bottom-up	Multi/Post-discipline	New explanation
Nanotechnology Top-down	Top-down	Top-down	Multi-profession	New product
Nanotechnology Bottom-up	Top-down	Bottom-up	Post-discipline	Process complexity

Note: A schematic listing of the general distinguishing characteristics of groups of practitioners engaged in various forms of nanotechnology. For instance, the aim of science is to proceed from first principles (bottom), and proceed (up) to increasingly sophisticated explanations of phenomena. The outcome is an explanation (of why it is so). Scientists generally identify themselves as working within a discipline. However, scientific experiments generally proceed top-down, beginning with plumbing and electrical wiring (or digging a very large hole in the case of CERN), and proceeding to exotic electronics and super-computers. 'Post-disciplinarity' is used here to describe an activity and its practitioners that belong to an area where the notion of discipline has lost its meaning (and is generally considered to be a hindrance). Attention is also drawn to the distinctiveness of the groups of practitioners that are engaged in top-down and bottom-up nanotechnology.

The concept of nanotechnology is refined here by choosing it as being defined by the activities of three groups, that is, scientists, technologists, and nanotechnologists, each of which operates in different settings, has different values, and pursues somewhat different goals. It can be argued that the centre of gravity at the present time, of activities in the broad field of nanotechnology, should properly be considered as basic and applied nanoscience, where the focus is more on the investigation of new phenomena associated with nanoscale structures, than on the explanation of the phenomena, and ultimately on the incorporation of that explanation into a broader scientific basis.

The established sciences are firmly anchored in their respective disciplines, and the various brands of technologists owe their allegiances to their respective professional societies. Nanotechnology, on the other hand, does not have such a disciplinary or professional home. Its practitioners are equally likely to be found in the departments of both biology and physics; indeed, research in nanobiology can quite happily be at home in a department of physics. There is a growing trend towards accommodating nanotechnology under a single institutional heading, where post-disciplinarity can be given free reign. In due course, it may be that institutionalized nanotechnology will follow the path of computing to become a discipline in its own right, and ultimately to divide into sub-disciplines (e.g., bio-nanotechnology, biomimetics, etc.).

1.4 MATERIALS CHARACTERIZATION—WHAT IS IT?

Materials characterization is a body of knowledge, and a general procedure, that allows us to deal constructively and rationally with questions such as:

- What are the actual attributes of a given material?

- Are the attributes of the material those that were intended, were promised, or were expected?
- Is a chosen materials processing route, or device fabrication route, working as intended, or as it did yesterday?
- What is different between the outcomes of two materials processing routes? Or why is the outcome different when it is believed that the same procedure is being followed?

In general, materials characterization is not concerned directly with the collection of information that is intrinsic, such as fundamental properties. However, the information arising from a particular characterization procedure may give rise to a set of data that can be used to determine properties. Thus, the term 'attribute(s)' has been chosen to describe the objective(s) of characterization.

As always, the secret is to ask the right questions, such as:

- Can a working hypothesis be formulated?
- What information is needed in order to confirm or deny the hypothesis?
- What is the 'best' way to obtain that information, by the most direct, reliable, and cost-effective route?
- Will the information aid, or obstruct, the advancement of present and future technologies?

Materials characterization may be carried out in various contexts, such as:

- When a project is at the basic, strategic, or applied stage.
- When a product or process is at the development stage.
- For purposes of process validation.
- During quality control.
- When troubleshooting is required.
- As a means of reverse engineering a process or product.

The first two above are concerned essentially with the transition from science to technology, and with the optimising of a product or process. In the third, the emphasis is on demonstrating the consistency, reliability, and robustness of the process. Quality control is usually concerned with obtaining a yes/no answer. Troubleshooting is usually needed when something goes wrong with an established procedure. The aim of reverse engineering is to determine retrospectively how a product has been made. The objective is then to ascertain whether the product has been made to specification. Alternatively, the objective might be to investigate a similar product from another source (such as a competitor).

1.4.1 STRATEGY VERSUS TACTICS OF MATERIALS CHARACTERIZATION

Strategy: Is concerned with the overall plan of approach in dealing with questions that can be answered wholly or in part by materials characterization.

Tactics: Is concerned with the detailed implementation and execution of a particular strategy.

Characterization will begin, implicitly or explicitly, with some thought being given to strategy and the general issues that are relevant to the task at hand. Once a strategy has been decided, the focus will then be on tactical issues.

There are several interrelated factors that will inform the thinking about materials characterization, and there are also several different entry points to the realm of strategy, each of which will lead to a different tactical route.

Materials possess *attributes*. The need to obtain information about a particular attribute will determine the choice of technique and method. For instance, compositional information cannot be obtained by probe microscopy, while on the other hand fluorescence spectroscopy will not provide insight into topography.

A particular material can be assigned to a *class* that endows it with certain general properties; examples of class include metal, insulator, polymer, composite, and so forth. On the nanoscale, a new class may be defined, in which characteristics and properties are size-dependent (for example, quantum effects). The class of a material will have a bearing on strategic thinking. For instance, techniques that are based on excitation by, or detection of, charged particles would not be suitable for insulators, or might require special procedures, while on the nanoscale one might expect unusual crystal symmetries and electronic properties.

Materials can be categorised by their *size, dimensionality, and extent of ordering*. 'Dimensionality' will be used to discriminate between objects that are 3-D, 2-D, 1-D, or indeed 0-D, where the latter refers to objects such as quantum dots and nanoparticles. Geometry and the extent of ordering are factors that will affect the choice of technique, operational mode, and method(s). In particular, the 'information volume' of the technique must be consistent with the characteristic length scale over which an attribute may change.

Sometimes, it is useful to enter via a particular 'technique', in which case it would be useful to know whether a favourite technique, or the one that is readily available, will indeed be able to deliver the required information.

1.4.1.1 Meso-Versus Nano-scale

There is no general agreement as to the location of the transition from the micro- to the nano-scale regime. It is useful to use 'meso' as a term for the grey area between the two regimes, since many current technologies can certainly operate in the sub-micro region, but are not yet truly nanoscale. It is also the case that 'traditional' characterization methods are increasingly being pressed into service on the mesoscale. Thus, some of the techniques and methods discussed in this volume will be relevant to the mesoscale rather than the nanoscale.

1.4.1.2 Dimensionality and Size Regimes of Meso/Nanostructures

There are two broad categories of meso/nano structures:

1. Nanostructured surfaces or ensembles with macro extension.
2. Single meso/nanoscale structures (of which there may be any number, essentially isolated and uncorrelated).

In the latter case, it is the size and dimensionality that broadly inform the strategic and tactical thinking about characterization.

1.4.1.3 Dimensionality

2-D: One dimension is nanoscale, and the other two are at least mesoscale, but more commonly macro/microscale. Characterization then reverts to the 'traditional' method(s) of thin film, surface, and interface analysis/characterization (e.g., Briggs and Grant 2003; Rivière and Myhra 2009). However, graphene and other monolayer structures merit consideration under the nanoheading, even though single objects may have microextent in two dimensions (see Chapter 11).

1-D: Two dimensions are nanoscale, while the third is micro/macroscale. Examples include nanowires, nanotubes, and biomolecular strands (DNA, microtubules, actin filaments, etc.); all are extremely topical and are attracting both scientific and technical attention (see Chapters 9 and 10).

0-D: All three dimensions are nanoscale. Examples include nanoparticles and single quantum wells; both are extremely topical. While the study of quantum wells is motivated principally by science, nanoparticles have important and widespread applications, often by virtue of having a large surface area-to-volume ratio (see Chapters 7 and 8).

There are micro- or meso-scale 2-D structures whose surfaces are ordered or para-ordered on the nanoscale. Examples include arrays of quantum wells, phase-separated polymers, zeolites, biomembranes, self-assembled monolayers, Langmuir–Blodgett films, etc. The problem in characterization then arises from the need to describe lateral differentiation, and structure, on the nanoscale, even though the surface/interface is macro/micro in lateral extent.

A particular attribute of a material may constitute a suitable entry point to the formulation of a strategy for materials characterization. In this case, such questions should be asked as:

- What is the morphology and/or the topography of the object?
- What is the crystal structure of the object or of its constituent sub-structures? Does it have long or short-range order? Is it amorphous? Is the surface/interface structure different from that of the bulk structure?
- What is the phase structure (if any)? Is it single phase or multi-phase? Which phases are present? Is it necessary to identify minor phases?
- What is the atomic or molecular composition? Is the object laterally or vertically differentiated in composition? Is trace element analysis required?
- What is the bulk/surface/interface chemistry? Is the surface chemistry differentiated laterally or in depth?
- What is/are the defect structure(s) (if any)? Do planar, line, or point defects need to be considered?

In some cases, detailed knowledge of the electronic structure will be required. Such experimental and theoretical studies go beyond what is normally regarded as materials characterization. (The use of the term 'chemistry' above is meant to imply routine determination of the chemical environment of a particular constituent species, which would thus be a guide to the molecular structure.)

There are also generic, but important, questions regarding

- The presence and effect of *contamination;*
- The *intrusiveness and/or destructiveness* of the probe;
- The effect of *specimen preparation;*
- The effect and role of the *substrate*; and
- The response of the structure to the *analytical environment* (e.g., *in situ* versus *ex situ*).

1.4.2 Macro-/Micro-scale Versus Nano-scale Materials Characterization

The techniques and methods that can be applied to materials characterization on the macro- and microscales are ubiquitous and well-established, and they have been the subjects of exhaustive description in the literature (see Brundle et al. 1992; Vickerman and Gilmore 2007; and Zhang, Li, and Kumar 2008; Yang and Li 2008 for recent examples). In most cases 'best practice', as specified by ISO standards, has been defined, and criteria for quality assurance laid down. For materials that are nanoscale in one dimension, that is, surfaces and interfaces, the state of affairs is equally satisfactory.

In the cases of systems that are nanoscale in two or three dimensions, the situation is quite different. There are certainly 'traditional' techniques, (e.g., particle sizing, X-ray diffraction (XRD), Raman, etc.), whose ranges of utility and operational capability have been extended, and which are now giving good service as nanotechnological tools, as described in Chapter 4 and Chapter 6, albeit principally as the means of obtaining information about ensemble averages. Other relatively 'old' techniques, such as transmission electron microscopy (TEM) and scanning electron microscopy (SEM), and their analytical attachments, have undergone several stages of technical and interpretational evolution, and are now genuinely nanotechnological characterization tools, described in Chapters 2 and 3, (e.g., Goldstein et al. 2003; Williams and Carter 2009; Yao and Wang 2008). Arguably, the most noteworthy relative newcomer is the large family of scanning probe microscopies (SPM) (e.g., Wiesendanger 1994; Bhushan 2010; Bonnell 2001; Rickerby, Valdre and Valdre 1990; Vilarinho, Rosenwaks, and Kingon 2005), see Chapter 5. (It is worth noting in passing that SPM has already delivered manipulation on an atom-by-atom scale, which was one of the requirements specified by Richard Feynman, but that development could be the topic of another monograph.) The electron-optical and scanning probe techniques can collectively deliver information volumes from which morphological, compositional, structural, and spectroscopic information can be obtained, on the meso- and nano-scales, for systems that are nano in two or three dimensions.

Importantly, from the point of view of this monograph, the literature seems to be relatively silent on a holistic description of how 'new' and 'old' techniques and methods can be brought to bear together on the interrogation of nanotechnology systems of great current interest.

Our assessment is that the current and near-future centre-of-gravity of nanotechnology can be found somewhere in the research and development phase. While this volume will address principally the needs in the research laboratory, it will not be entirely silent on the needs of quality control, industrial trouble-shooting, or online analysis. It will limit itself to techniques that are likely to be found in a research laboratory, rather than at large national facilities. Thus, techniques that are based on synchrotron or high-flux neutron sources will not be covered in any depth even though they are increasingly becoming available to a broad spectrum of users from public and private institutions.

1.5 CURRENT STATE OF 'BEST PRACTICE' AND QA

It is early days for definitive decisions on what constitute best practices and standards in nanotechnology. However, work is underway by committees and organisations in all the major industrial nations (e.g., BSI 2007; Carneiro et al. 2009; Danzebrink et al. 2006), and it is likely that a consensus will emerge in the next few years.

REFERENCES AND USEFUL READING

Bhushan, B., 2010, *Scanning Probe Microscopy in Nanoscience and Nanotechnology*, Berlin: Springer-Verlag.

Bonnell, D. A. (ed.), 2001, *Scanning Probe Microscopy: Theory, Techniques and Applications*, 2nd ed., New York, NY: Wiley-VCH.

Briggs, D. and Grant, J. T. (eds.), 2003, *Surface Analysis by Auger and X-Ray Photoelectron Spectroscopy*, Chichester: IM Publications.

Brundle, C. R., Wison, L., Evans, C. A., and Wilson, S., 1992, *Encyclopedia of Materials Characterization*, Stoneham, MA: Elsevier.

BSI, PD 6699-1, 2007, *Nanotechnologies—Part 1: Good Practice Guide for Specifying Manufactured Nanomaterials*, British Standards Institution.

Carneiro, K., Koenders, L., Ulm, G., Alcorta, J., Barbier, B., Eberhardt, W., Garnaes, J., Gee, M., Hatto, P., Hennecke, M., Long, G. G., Scholze, F., Van de Voorde, M. H., and Wilkening, G. 2009, Metrology, Standardization, Instrumentation: The Need for Nano-Metrology Research and Technology, *www.mf.mpg.de/mpg/.../pdf/02.../GENNESYS_2009-Chap6. pdf,* Max Planck Gesellschat.

Christensen, C. M., 1997, *The Innovator's Dilemma: When New Technologies Cause Great Firms To Fail*, Boston, MA: Harvard Business School Press.

Danzebrink, H. U., Koenders, L., Wilkening, G., Yacoot, A., and Kunzmann, H., 2006, CIRP Annals, *Manufacturing Technology,* 55, 841.

Drexler, K. E., 1992, *Nanosystems: Molecular Machinery, Manufacturing, and Computation*, Hoboken, NJ: Wiley Interscience.

Goldstein, J., Newbury, D. E., Joy, D. C., Lyman, C. E., Echlin, P., Lifshin, E., Sawyer, L. and Michael, J. R., 2003, *Scanning Electron Microscopy and X-Ray Microanalysis*, 3rd ed., Berlin: Springer.

Moore, G. E., 1965, *Electronics*, 38(8), 4.

Rickerby, D. G., Valdre, G., and Valdre, U. (eds.), 1999, Impact of Electron and Scanning Probe Microscopy on Materials Research, *NATO Science* Series E, Vol. 364.

Rivière, J. C. and Myhra, S. (eds.), 2009, *Handbook of Surface and Interface Analysis: Methods for Problem-Solving*, 2nd ed., New York, NY: CRC Press.

Stained Glass Association of America, 2010, History of Stained Glass. http://www.stainedglass. org/html/SGAAhistorySG.htm.

Vickerman, J. C. and Gilmore, I., 2007, *Surface Analysis*, 2nd ed., Hoboken, NJ: Wiley.

Vilarinho, P. M., Rosenwaks, Y., and Kingon, A., 2005, *Scanning Probe Microscopy, Characterization, Nanofabrication and Device Application*, New York: Springer-Verlag.

Walters, G. and Parkin, I. P., 2009, *J. Mater. Chem.*, 19, 574.

Wiesendanger, R., 1994, *Scanning Probe Microscopy and Spectroscopy: Methods and Applications,* Cambridge: Cambridge University Press.

Williams, D. B. and Carter, C. B., 2009, *Transmission Electron Microscopy: A Textbook for Materials Science*, Berlin: Springer.

Yang, F. and Li, J. C. M. 2008, *Micro and Nano Mechanical Testing of Materials and Devices*, Berlin: Springer.

Yao, N. and Wang, Z. L., 2004, *Handbook of Microscopy for Nanotechnology*, Berlin: Springer.

Zhang, S., Li, L., and Kumar, A., 2008, *Materials Characterization Techniques*, Boca Raton, FL: CRC Press.

BIBILIOGRAPHY

The following is a partial listing of recent and forthcoming monographs on nanoscience and nanotechnology.

Ahmed, W. and Jackson, M. J., 2009, *Emerging Nanotechnologies for Manufacturing*, Elsevier.

Balzani, V., Credi, A. and Venturi, M., 2008, *Molecular Devices and Machines*, Wiley.

Bhushan, B., 2010, *Springer Handbook of Nanotechnology*, p. 270, Springer.

Binns, C., 2007, *Introduction to Nanoscience and Nanotechnology,* Wiley.

Booker, R. and Feltman, E. B., 2005, *Nanotechnology for Dummies*, Wiley.

Boucher, P. M., 2008, *Nanotechnology*, Taylor & Francis.

Cao, G., 2004, *Nanostructures and Nanomaterials*, Imperial College Press.

Chaudry, Q. and Castle, L., R., 2010, *Nanotechnologies in Food*, RSC.

Contescu, C. I., Schwarz, J.A. and Putyera, K. (eds), 2008, *Dekker Encyclopedia of Nanoscience and Nanotechnology*, Taylor & Francis.

Frankel, F. C. and Whitesides, G. M., 2009, *No Small Matter*, Harvard University Press.

Goddard, W.A., Iafrate, G. J., Lyshevski, S. E. and Brenner, D. W. (eds), 2007, *Handbook of Nanoscience, Engineering, and Technology*, Taylor & Francis.

Gogotsi, Y. (ed.), 2007, *Handbook of Nanomaterials*, Taylor & Francis.

Heiz, U. and Landman, U., 2007, *Nanocatalysis*, Berlin: Springer.

Hester, R. E. and Harrison, R. M., 2009, *Nanotechnology*, RSC.

Hornyak, G. L., Dutta, J. and Tibbals, H. F., 2008, *Introduction to Nanoscience and Nanotechnology*, Taylor & Francis.

Hu, H., 2009, *Nanocrystals*, Nova.

Jennings, C. H., 2009, *Nanotechnology in the USA*, Nova.

Kelsall, R., Hamley, R., and Geoghegan, M., 2005, *Nanoscale Science and Technology*, Wiley.

Knauth, P. and Schoonman, J., 2007, *Nanocomposites*, New York: Springer.

Kosal, M., 2009, *Nanotechnology for Chemical and Biological Defense*, Springer.

Korgel, B. A. and Yacaman, M. J., 2007, *Fundamentals of Nanotechnology*, Wiley.

Lindsay, S., 2009, *Introduction to Nanoscience*, Oxford University Press.

Lowell, S., Shields, J. and Thomas, M. A., 2004, *Characterization of Porous Solids and Powders*, New York: Springer.

Lu, A.H., Zhao, D. and Ying, W., 2009, *Nanocasting*, RSC.

Lukehart, C. M., 2008, *Nanomaterials*, Wiley.

Majoros, I. J. and Baker, J., 2007, *Dendrimer-Based Nanomedicine*, World Scientific.

Meyyappan, M., 2007, *Introduction to Nanotechnology for Scientists and Engineers*, Wiley.

Meyyappan, M. (ed.), 2004, *Carbon Nanotubes: Science and Applications*, Taylor & Francis.

Mitin, V., Sementsov, D., and Vagidov, N., 2010, *Quantum Mechanics for Nanostructures*, Cambridge University Press.

Narlikar, A. V. and Fu, J. J., 2010, *Oxford Handbook of Nanoscience and Technology*, Vols. 1, 2, and 3, Oxford University Press.

O'Connell, M. J. (ed.), 2006, *Carbon Nanotubes*, Taylor & Francis.

Ozin, G. A. and Arsenault, A. C., 2005, *Nanochemistry*, RSC.

Peterson, W. P., 2009, *Nanotechnology*, Nova.

Poole, C. P. Jr. and Owens, F. J., 2003, *Introduction to Nanotechnology*, Wiley.

Ramsden, J., 2009, *Applied Nanotechnology*, William Andrew.

Riaz, U. and Ashraf, S. M., 2010, *Nanostructured Conducting Polymers and Their Nanocomposites*, Nova.

Reisner, E. and Bronzino, D. E., 2009, *Bionanotechnology*, Taylor & Francis.

Rogers, B., Adams, J., and Pennathur, S., 2007, *Nanotechnology*, Taylor & Francis.

Rusop, M. and Soga, T., 2009, *Nanoscience and Nanotechnology*, AIP.

Sattler, K. D., 2010, *Handbook of Nanophysics*, Taylor & Francis.

Seal, S., 2008, *Functional Nanostructures*, New York: Springer.

Schodek, D. L., Ferreira, P. and Ashby, M., 2009, *Nanomaterials, Nanotechnologies and Design*, Elsevier.

Shah, M. A. and Ahmad, T., 2010, *Principles of Nanoscience and Nanotechnology*, Alpha Science.

Sharma, K. R., 2009, *Nanostructuring Operations in Nanoscale Science and Engineering*, McGraw-Hill.

Shoseyov, O. and Levy, I., 2007, *Nanobiotechnology*, Humana Press.

Spowage, A., 2011, *Characterization of Nanostructured Bulk Materials*, Wiley.

Starov, V. M., 2010, *Nanoscience*, Taylor & Francis.

Terrazas, P. S., 2010, *Nanotechnology*, Nova.

Theodore, L., 2006, *Nanotechnology*, Wiley.

Theodore, L. and Kunz, R. G., 2005, *Nanotechnology*, Wiley.

Tibbals, H. F., 2010, *Medical Nanotechnology and Nanomedicine*, Taylor & Francis.

Tuan, V. D. (ed.), 2007, *Nanotechnology in Biology and Medicine*, Taylor & Francis.

Wee, A. T., (ed.), 2009, *Advances in Nanoscience and Nanotechnology*, World Scientific.

Wiederrecht, G., 2009, *Handbook of Nanofabrication*, Elsevier.

Wolf, E. L., 2006, *Nanophysics and Nanotechnology: An Introduction to Modern Concepts in Nanoscience*, 2nd ed., New York: Wiley-VCH.

Section I

Techniques and Methods

2 Electron-Optical Imaging of Nanostructures ((HR)TEM, STEM, and SEM)

2.1 INTRODUCTION

The quantum mechanical concept of particle-wave duality is the basis for electron microscopy. An electron wave with an energy of 300 keV, and with a de Broglie wavelength of ca. 2 pm, is the analogue of visible light with a wavelength of 500 nm. An electron wave has all the wave-like attributes such as phase, amplitude, coherence, luminosity, and direction of propagation. It might be thought that the point-to-point resolution of an electron microscope should therefore be of magnitude comparable to the electron wavelength, 2 pm, in accord with the diffraction limit, which is the limiting factor for optical microscopy. Unfortunately, electromagnetic lens aberrations conspire to degrade the diffraction-limited resolution by a factor of 10^2, to 0.1 nm, for the latest generation of aberration-corrected transmission electron microscopy (TEM) instruments. There are other similarities between the imaging mechanisms for optical and electron microscopy, such as diffuse scattering contrast (known as 'mass thickness contrast' for TEM) in transmission and phase contrast. On the other hand, there are some differences. For instance, a TEM cannot generate contrast in the backscatter mode, (but backscatter imaging is nevertheless very useful and popular in scanning electron microscopy (SEM)). Since the scattering mechanisms for electrons are different from those of photons, and because the wavelengths of keV electrons are comparable to interatomic spacings, spatially resolved structural information can be obtained readily in the diffraction mode.

A variation on the transmission theme, and one reminiscent of SEM, is that of scanning transmission microscopy (STEM). In an STEM, a fine-focus electron beam is rastered across a thin sample section and the transmitted intensity is detected as a function of position in the x-y plane. The spatial resolution can be comparable to that of high-resolution transmission electron microscopy (HRTEM). Since the interaction volume for very thin sections contains only a small number of atoms, the analytical resolution for energy dispersive spectroscopy (EDS) can approach that of a single atom.

As for many techniques, a description of electron microscopy will be bedevilled by acronyms. A list of the definitions of acronyms has been compiled as an aid to the reader, and can be found in the appendix to this chapter.

The relatively strong interactions experienced by electrons during their passage through condensed matter, in comparison to those of X-rays and neutrons, give rise to strong scattering and absorption. Accordingly, high-spatial-resolution imaging and analytical analysis by TEM and STEM require thin specimens (generally <150 nm, and sometimes down to a single or a few monolayers). In 'traditional' areas of materials characterization by TEM or STEM, the requirement for thin foil specimens often involved difficult, lengthy, and onerous specimen preparation. In the case of nanostructures, on the other hand, the requirement for thinness is satisfied, by definition, usually without any specimen preparation.

TEM and STEM methods have played central roles in the discovery, description, and characterization of nanostructures, partly due to near-atomic interaction and information volumes, but also due to the richness of structural, analytical, and spectroscopic information that can be obtained.

SEM will generally be the first port of call, after routine optical imaging, during characterization of nanostructures. It is a far more user-friendly technique than TEM, and places a lower premium on specimen preparation. On the other hand, it is principally a tool for topographical visualization. The current generation of instruments offers a point-to-point resolution of ca. 1 nm under favourable conditions, and in the secondary electron imaging mode. While an SEM is often fitted with EDS as an analytical attachment, the interaction volume is in that case inconsistent with nanoscale analysis.

The main physical principles and a schematic description of technical implementation will be covered here, followed by an overview of the contrast mechanisms of the main operational modes. The emphasis will be on those features and aspects that are most relevant to nanostructural characterization. Needless to say, the field of TEM is now extremely mature, and the complete gamut of capabilities is within the ambit only of seasoned experts. The technical literature is extensive; a small subset of the most widely available monographs includes (Williams and Carter; 2009; Goldstein et al. 2003; Reimer and Kohl 2008; Spence 2005; Titchmarsh 2009).

2.2 TEM OVERVIEW

A schematic of a typical conventional TEM instrument is shown in Figure 2.1.

The electron source for the current generation of TEM instruments is likely to be of the 'cold' field emission variety. This has the advantage of very high luminosity (10^7 A/mm^2/steradian), low-energy spread (0.3–0.5 eV at FWHM), and a source size defined by the apex of a sharp tip (in the region of 10 nm). An associated advantage is that the beam emitted from a field emission source is highly spatially coherent, which is a prerequisite for high-resolution imaging. The disadvantage is that it must be operated in a vacuum environment of 10^{-9} Pa or better, in order to prevent build-up of contamination on the apex, which would degrade the typical tip lifetime of some 10^4 h. The small source size is due to the enhancement of the local E-field at the apex.

The condenser lenses can be adjusted to produce either a defocussed parallel beam for imaging or diffraction analysis, or a focussed convergent beam for EDS or electron energy loss spectroscopy (EELS) analysis, at the sample plane. The adjustable and selectable condenser aperture defines the semi-angle of the convergent beam.

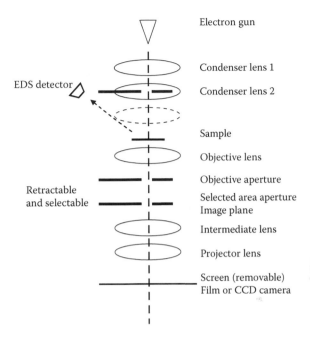

FIGURE 2.1 Schematic overview of a conventional TEM.

The major contribution to aberrations, which limit the resolution, arises in the objective lens. The objective aperture selects the central area of the lens thus minimizing spherical aberration, and can also be adjusted to select bright field (letting through the undeflected rays) or dark field (letting through a particular diffracted ray) diffraction contrast in the imaging mode.

A parallel beam is incident on the sample in the diffraction mode. The power of the objective lens is adjusted so that the undeflected rays are brought to a focus at the selected area aperture, located at the back focal plane. The diffracted rays interact with the straight-through rays to create a diffraction pattern in the image plane, where the pattern can be recorded (see Figure 2.2). The size and location of the aperture determine the particular area of the specimen from which a diffraction pattern will be obtained. In practice, the smallest area is ca. 0.5 μm. Accordingly, selected area diffraction (SAD) analysis is rarely useful for nanostructures.

2.2.1 MAGNETIC LENSES AND ABERRATIONS

A magnetic lens consists of a copper coil enclosed in a shaped soft iron pole piece, as shown in Figure 2.3. Magnetic field lines spread out from the gap. Accordingly, the electron beam will respond to field components that are longitudinal and transverse with respect to the direction of travel by the electron beam. The transverse Lorentz force acting on the beam is perpendicular to the plane defined by the direction of the magnetic field lines and the direction of travel. The transverse force components increase in strength away from the optical axis and bend, by an increasing amount, the electron rays towards the optical axis, analogous to the refraction of light rays by a thin optical lens.

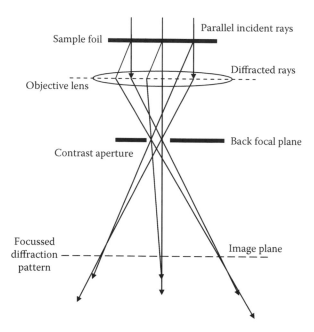

FIGURE 2.2 Conditions for the formation of a diffraction pattern in the image plane. (Adapted from Titchmarsh, 2009.)

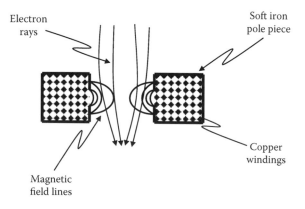

FIGURE 2.3 The principle of magnetic focussing.

The longitudinal force components give rise to a helical rotation of the beam. Good engineering and corrections can reduce astigmatic aberrations to near zero. Likewise, chromatic aberrations can be reduced by decreasing the energy spread of the electron beam. However, spherical aberration cannot be eliminated, and is always positive.

2.2.2 SPHERICAL ABERRATION

In the case of a perfect lens, all rays leaving a point in the object meet at one point in the image. If there is spherical aberration, then rays inclined to the lens axis arrive at an image point which is either closer to (positive aberration), or further from (negative aberration), the Gaussian image plane, than rays aligned with the axis. Electron lenses are electromagnetic and have positive aberration.

Rays inclined at an angle α to a lens axis form a disc of size $MC_s\alpha^3$, (M is the magnification), in the Gaussian image plane, where C_s is the spherical aberration coefficient. If two points in the image are a distance d_s apart, then each will produce a disc in the image of size $MC_s\alpha^3$, the centres being Md_s apart. These will overlap unless $Md_s > MC_s\alpha^3$. The limit of resolution due to spherical aberration is therefore $d_s = C_s\alpha^3$. More sophisticated treatments of circular aberration can be found in the literature (Ishizuka 1994; Spence 2003).

2.2.3 CHROMATIC ABERRATION

An electron beam will have a finite, although small, spread in energy, and thus in wavelength. Just as the focal length in photon optics depends on wavelength, so it does in electron optics. If the image plane for electrons of energy E is considered, then electrons of energy $(E + \Delta E)$ inclined at an angle α to the axis form a disc of size $MC_c(\Delta E/E)\alpha$, where C_c is the chromatic aberration coefficient; in other words, the limit of resolution due to chromatic aberration is $d_c = C_c(\Delta E/E)\alpha$.

2.2.4 RESOLUTION IN THE PRESENCE OF ABERRATIONS

A magnetic lens will suffer from spherical aberration and from the inevitable diffraction limit set by its size. The effects must add together in some way so as to produce the overall aberration. For simplicity, consider the case of a TEM, where α is small (and thus $\sin \alpha \approx \tan \alpha \approx \alpha$) and $n = 1$ (only the first-order diffraction effect is considered). Also, for algebraic simplicity, it will be assumed that the aberrations add arithmetically (analogous to a 'worst case' addition of errors rather than the 'likely situation' found by the addition of variances). That is, the resulting broadening in the focal plane, δ, is

$$\delta = \frac{0.6\lambda}{n\sin\alpha} + C_s\alpha^3 \qquad (2.1)$$

where
    λ is the de Broglie wavelength.

Differentiation with respect to α yields a minimum value for δ, giving a resolution of

$$\delta_{min} = (\lambda^3 C_s)^{1/4} \qquad (2.2)$$

This can be achieved with an optimum angular aperture of the lens α_{opt} of

$$\alpha_{opt} = \left(\frac{0.2\lambda}{C_s} \right)^{1/4} \qquad (2.3)$$

A typical 100 keV microscope may have $C_s = 1$ mm, with a de Broglie wavelength of $\lambda = 3.7$ pm, which gives

$$\delta_{min} \approx 0.5 \text{ nm} \quad \text{and} \quad \alpha_{opt} \approx 6 \times 10^{-3} \text{ radians}$$

For a focal length of 2 mm, this implies a physical aperture of diameter 25 μm in the objective lens. Thus, restricting the lens aperture does not always lead to a loss of resolution, as might be expected from the simple diffraction formula.

Chromatic aberrations are caused by the spread in wavelength, and thus energy, at the source and by an additional energy spread, adding up to ΔE, arising from inelastic interactions with the specimen (energy is always lost). The combination of chromatic aberration with diffraction effects gives for small α,

$$\delta = \frac{0.6\lambda}{\alpha} + C_c \alpha \frac{\Delta E}{E} \qquad (2.4)$$

Differentiation and rearrangement gives

$$\delta_{min} = 2 \left(0.6 C_c \lambda \frac{\Delta E}{E} \right)^{1/2} \qquad (2.5)$$

with

$$\alpha_{opt} = \left(\frac{0.6\lambda}{C_c} \times \frac{E}{\Delta E} \right)^{1/2} \qquad (2.6)$$

Taking $\Delta E = 1$ eV, $E = 100$ keV, $\lambda = 3.7$ pm and $C_c = 2$ mm gives

$$\delta_{min} \approx 0.4 \text{ nm and } \alpha_{opt} \approx 1 \times 10^{-2} \text{ radians}$$

The above are illustrative ballpark numbers. Current generation TEM instruments have point-to-point resolution better than 0.2 nm, and the very latest aberration-corrected models can offer somewhat below 0.1 nm.

2.3   INTERACTIONS OF ELECTRONS WITH MATTER

Interactions of energetic electrons with matter can conveniently be considered as either wave-like or particle-like processes. The former lead to descriptions of diffraction by periodic structures and to high-resolution imaging, while the latter lead to diffuse scattering and to analytical methods.

In the wave-like picture, an electron can be described as a travelling plane wave that evolves in space and time

$$\Psi = \exp[i(\omega t - \mathbf{k} \cdot \mathbf{r})] \tag{2.7}$$

Each electron is a wave packet of some nm in extent in the z-direction, which is much greater than the average separation between electrons. The transverse spacing, on the other hand, is consistent with the wave front being spatially coherent over the area of illumination. The latter is a requirement for high-resolution imaging and area diffraction analysis. The particle-like picture will be considered in more detail in Chapter 3.

Diffraction analysis by periodic structures is a powerful and widely used method for the investigation of those samples that are crystalline and of microscale dimensions along two axes. The underlying theory has origins dating back to the earliest descriptions of the visible part of the electromagnetic wave spectrum. X-ray diffraction analysis provided further impetus to the development of the theory, and electron-wave diffraction has adopted most of the formalism and concepts from that form of analysis. Accordingly, the well-known Bragg equation remains the principal tool for the indexing of a diffraction pattern, and thus the determination of interlayer spacings and relevant symmetries of crystal structures, viz.,

$$\lambda = 2d_{hkl} \sin \theta_{hkl} \tag{2.8}$$

where λ is the wavelength of the incident radiation, d the interplanar spacing, and θ the scattering angle. The Miller indices, hkl, denote a particular set of planes.

The lateral spatial resolution of selected area diffraction (SAD) is limited by the size of the aperture. In practice, the limit is ca. 0.5 µm. Thus, SAD analysis is rarely useful for nanostructural investigations. Rather than delving further into the topic here, the reader is referred to the literature (e.g., Cowley, 1975).

In the present context of the characterization of nanostructures, the widely used and well-established bright field and dark field imaging conditions, both of which are based on diffraction contrast, are of limited usefulness on the nanoscale. The general ideas are shown in Figure 2.4. The reader is referred to the literature for an in-depth discussion (Reimer and Kohl 2008; Williams and Carter 2009; Forwood and Clarebrough 1991).

2.3.1   HIGH-RESOLUTION IMAGE FORMATION

In a crystalline solid, the lattice potential is periodic and can be represented as a Fourier series. Near the atom cores, higher-order terms are required in order to describe the strong spatial dependence, while at the mid-points between neighbouring

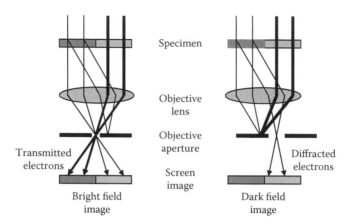

FIGURE 2.4   Schematic description of the ray-paths and the positions of the objective aperture for bright field (left) and dark field (right) imaging conditions.

atoms, the spatial dependence is more gentle, requiring a few lower-order Fourier terms for a good description. An electron travelling through a lattice will respond to the local potential, and the scattering angle will depend on the proximity to the atom core.

When a crystal is oriented with a zone axis parallel to the incident beam (a zone axis is defined by the line of intersection between two lattice planes), the beam then illuminates ordered columns of atoms, as shown in Figure 2.5. The columns are surrounded by corrugated equipotential contour planes. The equipotential planes nearest to the centres of the columns will have deep corrugations, while those that

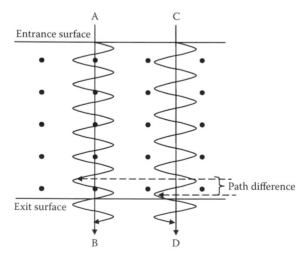

FIGURE 2.5   Two electron paths, A to B and C to D are shown, resulting in a path difference, due to the fact that the A to B path propagation is close to the atom column (solid dots), while that of C to D is between the columns (From Titchmarsh, 2009, with permission from Taylor & Francis.)

are midway between columns will have gentle corrugations. The electron wavelets emanating from each point of the broad plane-wave front of the incident electron beam will scatter, by diffraction, off the contour planes. The diffraction pattern then becomes a Fourier transform in two dimensions, of the periodic potential. The objective lens then brings the diffracted beams and the straight-through beams together to produce a phase-contrast image. The interference pattern is essentially a back-transformation and leads to an enlarged picture of the periodic potential.

It might be thought that, with an electron wavelength of 1.97 pm at 300 keV, it should be possible to reconstruct an image from Fourier components with a comparable wavelength resolution. Regrettably, several factors conspire to degrade the resolution by a factor of around 50, to 0.1 nm, for even the best instruments. The limitations arise from a combination of spherical and chromatic aberrations in the objective lens, of lack of complete spatial and temporal coherence in the illuminating beam, of the finite thickness of the sample, and of contributions from electronic and mechanical noise and drift. A further problem that makes the interpretation of high-resolution images difficult is a variable phase shift introduced by the objective lens; this shift depends on the magnification.

A somewhat more formal approach can be taken to illustrate the contrast mechanism, and the limits on resolution.

An incident electron can be described as a plane wave of unit amplitude (neglecting the time dependence), viz.,

$$\Psi_i = \exp(-i\mathbf{k} \cdot \mathbf{r}) \tag{2.9}$$

The transmitted wave, Ψ_t, will have undergone a phase change, leading to

$$\Psi_t = \exp(-i\mathbf{k} \cdot \mathbf{r})\exp(-i\sigma\phi_0 t) \tag{2.10}$$

where ϕ_0 is the average potential, and t is the thickness of the sample. The factor σ is equal to $2\pi me\lambda/h^2$, (λ and m are the wavelength and mass of the incident electrons, respectively). In the weak phase object approximation (WPOA), when the scattered intensity is much lower than that of the primary beam, the exponential phase factor can be expanded, and the two leading terms retained. Thus,

$$\exp(-i\sigma\phi_0 t) \approx 1 - i\sigma\phi_0 t$$

The *local* potential, ϕ_P, as opposed to the average potential, for the path A to B shown in Figure 2.5 will be different from that for the path C to D. Accordingly, there will be a relative phase difference with respect to the two paths, as well as an intensity change. The scattering term $\sigma\phi_P(r)$ thus contains all the information about the structure. The WPOA assumption becomes less valid as the scattered intensity increases with respect to the incident intensity. For low-Z materials, the limit of validity can be reached for a path length of a few nm. Resort has then to be made to computer modelling in order to validate an observed image.

The information contained in the diffraction pattern is transferred by the objective lens. The aberrations of that lens change the phase for each diffracted ray, with a particular diffracted angle, α, and a particular azimuthal angle, ϕ. The resulting distorted information emerging from the objective lens will, therefore, give rise to a distorted lattice image in the image plane, where the distortion is a function of α and ϕ (note that the symbol ϕ is used to denote potential as well as azimuthal angle).

The complex relationship between spherical and chromatic aberrations, defocus conditions, spatial frequency, and spatial coherence, can be expressed as a contrast transfer function (CTF) (Titchmarsh 2009), that is,

$$CTF = \exp[-\pi^2 q_0^2 (C_1 \lambda \mathbf{K} + C_3 \lambda^3 \mathbf{K}^3)^2 \sin[(2\pi / \lambda) d\chi(\mathbf{K}) / dK] \qquad (2.11)$$

$$\chi(\mathbf{K}) = [C_1 \lambda^2 \mathbf{K}^2 + C_3 \lambda^4 \mathbf{K}^4 / 2] / 2 \qquad (2.12)$$

Here, χ relates the phase distortion to the defocus (C_1 is the coefficient of defocus) and to the spherical aberration (C_3 is the coefficient of three-fold spherical aberration), as a function of scattering vector corresponding to a particular Bragg condition. Thus, the CTF is an oscillating (sine) function of the gradient of χ with respect to the scattering vector. The exponential term arises from the dependence on partial spatial coherence, the extent of which depends on the maximum semi-angle, q_0, of focussed illumination.

The sine term of the CTF is plotted in Figure 2.6. As to be expected, the function oscillates between +1 and –1. However, there is a 'pass-band' for Fourier components around 2 nm^{-1} (the implication of negative contrast is that atoms will appear dark on a bright background). The position and width of the band depend on the defocus, that is, the value of C_1. Within the pass-band, the exit wave function is transferred by the objective lens without spurious contrast reversal, and the image can be interpreted with some confidence. The crossover at zero occurs for k $\approx$ 4.7 nm^{-1}, corresponding to a point-to-point resolution of ca. 0.21 nm. The pass-bands at greater k

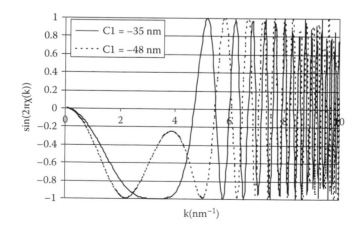

FIGURE 2.6 A plot of the sine term of CTF as a function of wave vector for two defocus values (C1 = -35 and -48 nm), and for q_0 = 1 mrad, C_3 = 0.5 mm, and E_0 = 200 keV (From Titchmarsh, 2009.)

reverse contrast for small changes in defocus, and will, in practice, be washed out. Additional discussion can be found in the literature (Spence 2003).

Other effects that affect the ultimate resolution can be considered under the heading of temporal coherence. The effects are due to instabilities and drift in the accelerating power supply, in the energy spread at the electron source, and in mechanical and electrical noise. Their combined effect is to modify the CTF function by an exponential multiplicative term that increasingly degrades the contrast with increasing wave vector. Good design and careful engineering is being applied to successive generations of instruments, in combination with advances in aberration correction, in order to broaden the pass-band.

In general, the procedure adopted for high-resolution imaging includes the recording of a series of images around the Schertzer defocus value, defined by $C_1 = (C_3\lambda)^{1/2}$ (Schertzer 1949), by incremental changes in C_1. The series of images is then compared to synthetic computed images. Recent developments in the field can be found in the literature, (e.g., Erni, Rossell, and Nakashima 2010).

An illustration of an atomically resolved high-resolution image of the interface region of $SrTiO_3/BaZrO_3$ bilayer thin films on a MgO substrate, is shown in Figure 2.7 (Mi et al., 2007).

2.4 ABERRATION CORRECTION

A typical aberration corrector consists of multiple elements. In the case of the Nion design for the VG Microscopes HB501 STEM (see Figure 2.8), there are four quadrupoles (red) and three octupoles (blue) placed between the two condenser lenses (bottom) and the objective lens (top) (Dellby et al. 2001). The quadrupoles shape the beam into a pencil-like cross section inside the octupoles. The octupoles then provide correction for the spherical aberration in the objective lens. The corrector optics have overall unity gain in both the x-z and y-z planes, as seen by comparing image sizes presented to the corrector and transferred to the top condenser lens, C2. Electrical and mechanical stabilities of 0.2 ppm are crucial to the successful operation of this device. The image on the right in Figure 2.8 shows the corrector hardware mounted underneath the microscope objective lens.

2.5 SCANNING TRANSMISSION ELECTRON MICROSCOPY (STEM)

Conventional TEM generates images by a parallel process, in the sense that the entire specimen area of interest is illuminated by a broad plane-wave beam, and each pixel in the image plane is integrated and counted in parallel. An alternative is to borrow from SEM technology, and to generate transmission images and spatially resolved analytical information by rastering a narrow convergent beam across a field of view of a thin specimen. The image acquisition time is then much longer due to serial counting. This alternative is known as scanning transmission electron microscopy (STEM). Many TEM instruments can be operated in the convergent beam mode (CTEM) as well as in the STEM mode. The former is rarely used for nanostructure imaging. The best STEM performance is obtained from dedicated instruments, in which spatially resolved analytical information (from energy dispersive X-ray

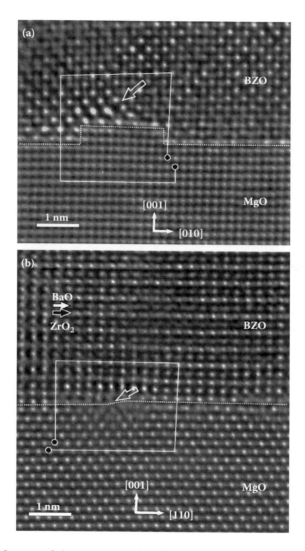

FIGURE 2.7 Images of the structure at interface areas between BZO and MgO viewed along (a) the [100] zone axis, and (b) along the [110] zone axis. The interfaces of BZO/MgO are marked by dotted lines. Arrows denote the dislocations at points where the interface structure changes from ZrO_2/MgO to BaO/MgO. (From Mi et al., 2007, *J. Cryst. Growth*, with permission from Elsevier.)

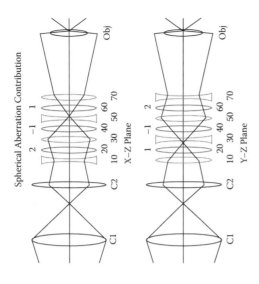

FIGURE 2.8   (See color insert.) The Nion-designed aberration corrector for an STEM instrument, consisting of a stack of quadrupole and octupole lenses between the objective and condenser lenses. (From IBM™ Web site, also see Dellby et al., 2001.)

spectroscopy (EDS) or electron energy loss spectroscopy (EELS)) can be obtained in parallel with high-resolution imaging, and can be of great assistance in interpreting atomically resolved images. The present discussion will refer to dedicated STEM instruments and their operational modes.

An STEM instrument can make use of several signals for imaging, including large-angle elastic scattering leading to (high-angle) annular dark field (ADF) and (HA-ADF) imaging, and small angle elastic scattering leading to bright field (BF) imaging. For analytical purposes, small-angle inelastic scattering allows electron energy loss spectroscopy (EELS), as well as the production of characteristic X-rays for energy dispersive X-ray spectroscopy (EDS). The analytical modes will be considered in Chapter 3.

2.5.1   TECHNICAL IMPLEMENTATION

A simplified schematic of an STEM instrument is shown in Figure 2.9. The top end of an STEM instrument is similar to that of a cold-cathode field emission SEM, although in an STEM, electrons are normally accelerated to energies of 60–100 keV, as opposed to the SEM energy range from <1 to >10 keV.

The incident beam is convergent, with a semi-angle q_0, (see Figure 2.10), and is focussed on the sample plane. The beam diameter at focus can be sub-nm (present state-of-the-art is less than 0.1 nm) (Batson, Dellby, and Krivanek 2002), and the monochromated energy linewidth can be less than 0.1 eV (Mook, Batson, and Kruit 2000), which has advantages for spatial resolution and more particularly for high spectral resolution in EELS analysis.

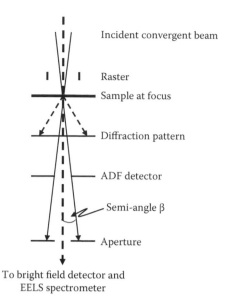

Incident convergent beam

Raster

Sample at focus

Diffraction pattern

ADF detector

Semi-angle β

Aperture

To bright field detector and
EELS spectrometer

FIGURE 2.9 Simplified schematic of an STEM instrument. EDS and EELS attachments are not shown. The aperture defines the semi-angle β for collection of low-order diffracted electrons that will contribute to the bright field image.

2.5.2 Diffraction Information from STEM

When the incident beam is convergent, the Bragg condition is satisfied for a range of angles equal to the angular spread of incidence. Accordingly, the diffraction pattern will consist of discs, rather than discrete points, which happens for a parallel incident beam. The situation is illustrated in Figure 2.10. For high Bragg angles, the diffraction discs will overlap. The range of overlap defines the regime in which low-order diffracted beams can interfere with the undeflected beams and thus degrade the bright field image. The image formation mechanism is similar to that of a high-resolution image in a TEM, and the outcome is subject to the same constraints in terms of defocus, aberrations, and instabilities (Titchmarsh 2009).

In the dark field imaging mode, electrons are collected by an annular ADF detector; the electrons forming the bright field image pass through the central hole, and are lost to the ADF detector, and the contrast is formed by the coherent interference of the diffracted beams. At greater scattering angles, the image becomes increasingly incoherent. However, the implication of the greater scattering angle is that the scattering process is due to electrons passing through the strong core potential near the atomic nucleus. This imaging mode is known as the HA-ADF. While the image intensity is low, the spatial resolution can be better than that of the ADF mode, and is independent of coherence. Accordingly, there is no contrast reversal with defocus and the image can often be interpreted directly. High-angle scattering depends on the atomic number of the scattering atoms, and the analysis of low-Z materials has been challenging. However, aberration correction has improved the situation a great deal in recent years, and atomically resolved images can now be obtained for low-Z

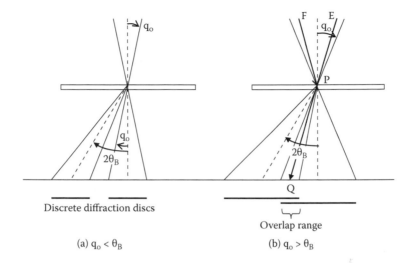

FIGURE 2.10   Illustration of the diffraction disc patterns obtained for a convergent incident beam. If the Bragg angle θ_B is greater than the semi-angle, q_0, of the convergent beam, then the pattern will consist of separated discs. If $q_0 > \theta_B$, then the discs will overlap. A ray from E can reach a point Q, passing through point P in the sample foil, without diffraction. A ray from F can be diffracted through $2\theta_B$ at P, and will also arrive at Q. Thus, there will be a mixing of diffracted and undiffracted beams, with a relative phase change. (From Titchmarsh, 2009, with permission from Taylor & Francis.)

nanostructures (see Chapter 9). The combination of STEM imaging with atomically resolved EDS and EELS analysis is a particular attractive feature, and will be described in Chapter 3.

2.6   THE ISSUE OF RADIATION DAMAGE DURING IMAGING AND ANALYSIS

The best resolution from the latest generation TEM instruments is generally achieved in the primary energy range 200–300 keV. Many of the most topical nanostructures consist wholly, or partly, of low-Z elements (e.g., C_{60}, carbon and BN nanotubes, graphene, functionalised structures, etc.). The energy transfer, T (in eV), for head-on collisions between relativistic electrons and atoms is given by

$$T = \frac{560.8}{A} \times \frac{E}{0.511}\left(\frac{E}{0.511} + 2\right) \tag{2.13}$$

where A is the atomic mass number, E is the electron kinetic energy in MeV, and 0.511 is the electronic rest mass in MeV. The energy transfer at 200 keV is thus about 40 eV, which is well above the threshold for displacement damage.

Specimen damage can be largely eliminated by operating the TEM at < 80 keV, which is accompanied by a relatively modest loss in resolution. In addition, it helps

to work with a low-luminosity beam during the survey and optimization period, at
the cost of grainy images.

2.7 SEM

Optical microscopy is generally the first port of call during a programme of nanoscale
topographical characterization, followed closely by SEM imaging in the secondary
electron mode.

2.7.1 TECHNICAL OVERVIEW

Scanning electron microscopy is the mainstream technique for imaging on dimensional scales from the macro- to the nano-regimes. The SEM is similar to the TEM
in that both employ a beam of electrons directed at the specimen. This means that
certain features, such as the electron gun, condenser lenses, and vacuum systems,
are similar in both instruments, although the ways in which the images are produced
are entirely different. Whereas the TEM provides information about the internal
structure of thin specimens, the SEM is used primarily to image the surface, or near-
surface, topography of bulk specimens. As well, SEM can image in the backscatter
mode which will reveal atomic number contrast, but at relatively modest lateral resolution, and SEM instruments can also be fitted with EDS analytical attachments. The
backscatter imaging mode and EDS analysis are largely irrelevant to nanostructural
characterization due to the relatively large interaction volumes (typically of 0.5 μm
depth and lateral extent).

Unlike in a TEM, the aberrations due to the electromagnetic optics manifest
themselves only in the quality of the incident probe beam that is being focussed
on the specimen surface by the objective and condenser lenses. The magnification
is defined by the ratio of the projected on-screen image divided by the size of the
raster. The point-to-point resolution is therefore a function of the spot size, which is
dependent on the diameter of the effective electron source, and on aberrations in the
optics. Typical figures of merit for the current state-of-the-art for a cold-cathode field
emission source are summarised in Table 2.1.

2.7.2 COLD-CATHODE FIELD EMISSION

Field emission from a cold flat cathode surface depends on the work function, (the
energy required to take an electron from the highest occupied state in the surface
to a state at infinity), and the local electrostatic field, referenced to an anode some
distance away. A typical work function is ca. 4.5 eV. This is the minimum energy
required for 'classical' electron emission from a surface. However, electrons can
tunnel through the surface barrier; thus emission will usually commence at lower
energies, and is generally described by the Fowler–Nordheim tunnelling processes.
The local electrostatic field, for a given potential difference between anode and cathode, depends on the local geometry of the cathode. In particular, if the cathode has
the shape of a needle with a small radius of curvature, r, the field at the apex of the
needle is enhanced by a factor that can be greater than 10^3, in comparison with the

TABLE 2.1

Cold-Cathode Field Effect Emitter Source

Operating temperature [K]	300
Cathode radius [nm]	< 100
Effective source radius [nm]	2.5
Emission current density [A/cm²]	17,000
Total emission current [µA]	5
Normalised brightness [A/cm² · sr · kV]	2×10^7
Maximum probe current [nA]	0.2
Energy spread at cathode [eV]	0.26
Energy spread at gun exit [eV]	0.3–0.7
Beam noise [%]	5–10
Emission current drift [%/h]	5
Operating vacuum [mbar]	$< 1 \times 10^{-10}$
Typical cathode life [h]	> 2000

Note: Typical figures of merit for a cold-cathode field emission electron source. Particularly noteworthy figures are those of the source radius, the energy spread (and thus the effect on chromatic aberration), and the brightness/luminosity. Lifetime is also an important practical factor.

field in the space between parallel plates. A useful approximation, widely used in the literature (e.g., Sadeghian and Kahrizi 2008), to the local field in the vicinity of the apex of a sharp tip, is

$$E_{local} = V_0/(kr) \tag{2.14}$$

where k is the so-called field factor, ranging from 3 to 8 and V_0 is the cathode to anode potential. For an accelerating voltage of 5 kV, and a radius of curvature at the apex of the cathode emitter of 10 nm, E_{local} may be of order 10^8 V/m.

A schematic of an SEM instrument is shown in Figure 2.11.

2.7.3 POINT-TO-POINT RESOLUTION

The brightness, β, of the electron source is given by

$$\beta = \frac{4I}{(\pi d_0^2)(\pi \alpha_0^2)} \tag{2.15}$$

where I is the current incident on the anode aperture, of diameter d_0, and α_0 is the angle of primary beam divergence. The final beam diameter, d_3, defined by the final aperture, usually the third, incident on the sample, is given by $d_3 = d_0 M_{total}$, where

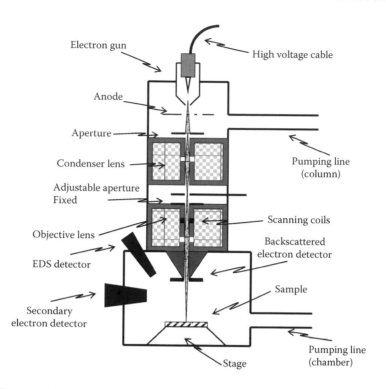

FIGURE 2.11 Schematic of a typical SEM showing the optical path, EDS attachment, and sample stage. The principles and practice of EDS will be discussed in Chapter 3. The main function of the fixed apertures is to clean up the primary by removing electrons that have been scattered out of the primary central part of the Gaussian profile distribution. The sample stage usually allows x-y translation, as well as rotation and tilt, of the sample.

M_{total} is the overall combined demagnification of the column optics (where lens aberrations have been neglected). The actual beam diameter, d_s, will be degraded by spherical lens aberration, leading to an expression of the form

$$d_s = a \left(b \frac{i_B}{\beta} + \frac{c}{V} \right)^{3/8} C_s^{1/4} \tag{2.16}$$

where a, b, and c are constants, i_B is the total current passing through the anode aperture, and C_s is the coefficient of spherical aberration. Since the objective lens is between the final aperture and the sample, it is apparent that $d_s > d_3$. As well, from Equation 2.16, it can be seen that resolution will be degraded for lower luminosity and acceleration voltage.

2.7.4 DEPTH OF FIELD

By virtue of the relatively much smaller angle of convergence of the incident beam, in comparison to that in optical microscopy, the depth of field, that is, depth of focus,

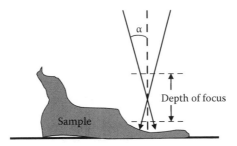

FIGURE 2.12   Illustration of the depth of focus in secondary electron imaging with an SEM. Being dependent on the angle of convergence of the incident beam, the depth of focus must depend on the working distance.

is much greater for an SEM than for the optical equivalent. The sketch in Figure 2.12 illustrates the point. While for visible confocal optics, the depth of focus is comparable to the lateral spatial resolution, simple scaling of the respective angles of convergence of the incident beam gives SEM a depth of focus greater than the point-to-point resolution by a factor of 10^3, or more. Obviously, the depth of focus is related to the working distance, and requires judgements to be made about the acceptable degradation in point-to-point resolution above and below the focal point.

2.7.5   IMAGING MODES

The focused incident electron beam, of energy 0.5–30 keV, is rastered over the surface of a sample by a scan generator, in the same manner as that in which a TV picture is generated. The energetic electrons interact with the sample, and give rise to a number of processes, the two principal ones of which, for the generation of image contrast are:

- The emission of low-energy secondary electrons ejected from valence and conduction states in the surface or near-surface.
- The backscattering of incident electrons, where the backscattered intensity depends strongly on the average atomic number Z of the atoms within the interaction volume.

Both processes can be used as means for the production of an image, either a secondary electron (SE) image which reveals surface topography, or that from backscattered electrons (BSE), which reveals lateral variations in near-surface composition, to a depth of 0.5–1 µm, and with a comparable lateral resolution. However, the latter imaging mode is largely irrelevant for nanoscale characterization, due to the µm-scale interaction volume.

2.7.5.1   Secondary Electron Imaging

Secondary electrons are detected by a system consisting of a scintillator backed by a photomultiplier. The secondary electrons are attracted to the detector by a bias voltage of 0–200 V on a grid. They are then accelerated onto the scintillator (a phosphor), which emits light (the reverse photo-electric effect). That light is transmitted via a light pipe to a photomultiplier, which converts the photons into an amplified pulse of electrons.

The secondary electron yield is also sensitive to the local voltage of the specimen, where the specimen might be an electrical or electronic network under power, thus presenting a surface with laterally differentiated potential. This is the basis of the 'voltage contrast' imaging mode that is used to investigate working electronic devices.

The same scan generator that is used to drive the deflector coils is also used to scan the photocurrent spot across the video screen, the brightness of the spot being modulated by the amplified signal from the detector. The electron beam and the spot on the screen are both scanned in a way similar to that in a television receiver, that is, in a rectangular set of parallel lines known as a 'raster'. The ultimate magnification of an SEM is the ratio of the linear distance, L, between two resolvable points on the display screen divided by the point-to-point resolution of the instrument. Unlike in a TEM, lens aberrations play no part in the projection of the image

2.8   EXAMPLES OF SEM PERFORMANCE

Two examples of high-resolution images obtained with the latest generation of SEM instruments are shown in Figures 2.13 and 2.14.

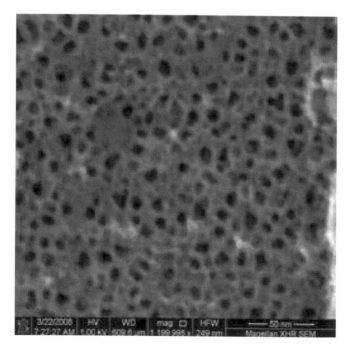

FIGURE 2.13 Illustration of high-resolution SEM imaging of a poly-silicon surface obtained with a Magellan SEM. The field of view is 250x250 nm^2; the average diameter of the resolved features is around 10 nm. (From http://www.fei.com/uploadedFiles/Documents/Content/Magellan-applicationBrief.pdf.)

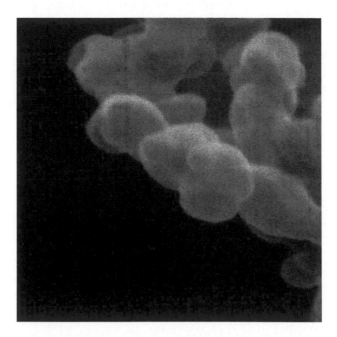

FIGURE 2.14 High-resolution SEM image of graphitised carbon, obtained with an ORION® PLUS instrument. The field of view is 100×100 nm^2. (From Scipioni, L., 2008, Carl Zeiss, NTS, SMT Web site, application note. With permission.)

2.9 OPTIMIZATION OF IMAGE QUALITY

SEM is a flexible, convenient, and user-friendly means for nanoscale topographical characterization. Images are readily interpretable by non-experts. Nevertheless, there are some issues that need to be considered in order to achieve the best results, including choices of aperture, working distance, and procedures for dealing with insulating materials. The effects of these choices are shown schematically in Figures 2.15 and 2.16.

2.9.1 INSULATING MATERIALS

Insulating samples can present problems due to charging. In extreme cases, and for the earlier generations of instruments, it has been common practice to sputter-coat the specimen with a thin layer of conductive material in order to obtain usable results. Instruments with field emission sources operate at relatively low primary energies, typically 1–2 keV, and low beam currents, in the range of 10 pA. Accordingly, moderate insulators, such as doped semiconductors, can be imaged satisfactorily without resorting to coating. The situation can be improved further by decreasing the energy to less than 1 keV, albeit with a loss of resolution by a factor of about 2.

ACKNOWLEDGEMENTS

We appreciate helpful comments and conversations with Dr. K. Jurkschat and J. Critchell.

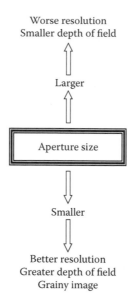

FIGURE 2.15 The effects on image quality of size of aperture. A larger (smaller) aperture provides higher (lower) beam current, but more (less) aberration, and a greater (smaller) angle of convergence.

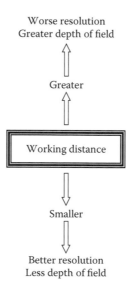

FIGURE 2.16 The effects of working distance on image quality.

APPENDIX: DEFINITIONS OF ACRONYMS USED WIDELY FOR DESCRIPTION OF ELECTRON MICROSCOPY (IN ALPHABETICAL ORDER)

ADF: Annular Dark Field (detector/detection)
BSE: Backscattered Electron (image/imaging mode)
CTEM: Convergent Beam TEM
CTF: Contrast Transfer Function
EDS: Energy Dispersive Spectroscopy (also known as EDX or EDXS)
EELS: Electron Energy Loss Spectroscopy
HA-ADF: High Angle ADF
(HR)TEM: (High Resolution) Transmission Electron Microscopy
SAD: Selected Area Diffraction
SE: Secondary Electron (image/imaging mode)
SEM: Scanning Electron Microscopy
STEM: Scanning Transmission Electron Microscopy
WPOA: Weak Phase Object Approximation

REFERENCES

Batson, P. E., Dellby, N., and Krivanek, O. L., 2002, *Nature* 418, 617.

Cowley, J. M., 1975, *Diffraction Physics*, Amsterdam, The Netherlands: Elsevier Science B.V.

Dellby, N., Krivanek, O. L., Nellist, P. D., Batson, P. E., and Lupini, A. R., 2001, *J. Electron Microscopy* 50, 17.

Erni, R., Rossell, M. D., and Nakashima, P. N. H., 2010, *Ultramicroscopy*, 110, 151.

FEI, http://www.fei.com/uploadedFiles/Documents/Content/Magellan-applicationBrief.pdf.

Forwood, C. T. and Clarebrough, L. M., 1991, *Electron Microscopy of Interfaces in Metals and Alloys*, New York, NY: Adam Hilger.

Goldstein, J. I., Lyman, C. E., Newbury, D. E., Lifshin, E., Echlin, P., Sawyer, L., Joy, D. C., and Michael, J. R., 2003, *Scanning Electron Microscopy and X-Ray Microanalysis*, 3rd ed., New York: Springer.

IBM, http://domino.research.ibm.com/comm/research_projects.nsf/pages/stem_eels.corrector.html.

Ishizuka, K., 1994, *Ultramicroscopy*, 55, 407.

Mi, S. B., Jia, C. L., Faley, M. I., Poppe, U., and Urban, K., 2007, *J. Cryst. Growth*, 300, 478.

Mook, H. W., Batson, P. E., and Kruit, P., 2000, *Proc. 12th Eur. Cong. on Electron Microsc.*, Vol. III, 315.

Reimer, L. and Kohl, H., 2008, *Transmission Electron Microscopy: Physics of Image Formation*, 5th ed., New York, NY: Springer.

Sadeghian, R.B. and Kahrizi, M. 2008. *IEEE Sensors J.*, 8, 161.

Schertzer, O., 1949, *J. Appl. Phys.*, 20, 20.

Scipioni, L., 2008, Carl Zeiss SMT Web site, application note, http://download.zeiss.de/nts/lifescience/PI_Ultra-High_Resolution_Imaging_in_ORION_PLUS.pdf.

Spence, J. C. H., 2003, *High Resolution Electron Microscopy*, 3rd ed., Oxford, UK: Oxford University Press.

Titchmarsh, J., 2009, In *Handbook of Surface and Interface Analysis: Methods for Problem-Solving*, 2nd ed., Rivière, J. C. and Myhra, S., (eds.), Boca Raton, FL: CRC Press.

Williams, D. B. and Carter, C. B., 2009, *Transmission Electron Microscopy: A Textbook for Materials Science*, 2nd ed., New York, NY: Springer.

3 Electron-Optical Analytical Techniques

3.1 INTRODUCTION

Imaging and structural analysis of nanostructures by electron-optical microscopy were considered in Chapter 2 in the context of the wave-like character of electrons. In this chapter, the particle-like character of electrons will be more appropriate for a discussion of analytical techniques that are commonly associated with electron microscopy. In particular, it will be seen that energy dispersive spectroscopy (EDS) and electron energy loss spectroscopy (EELS) can offer information on composition and chemistry (i.e., electronic structure) with near-atomic resolution. There are several excellent texts that describe the principles and practice of EDS (Russ 1984; Williams, Goldstein, and Newbury 1995; Goldstein et al. 1992), although the emphasis in the literature has tended to be on EDS in combination with SEM, or on WDS (wave-length dispersive spectroscopy) as carried out with an electron microprobe instrument. Here the emphasis will be on EDS in combination with TEM and STEM. The monograph literature on EELS is rather more sparse (Brydson 2001; Ahn 2004; Egerton 2009).

In EDS, an incident energetic electron scatters inelastically during its passage through the core potential of an atom, thereby ejecting an electron from a core state, leaving the atom in an excited state. The return to its ground state can then result in the emission of an X-ray quantum with energy characteristic of a core level transition. For EELS, it is the many types of scattering process, each one giving rise to a characteristic energy loss, which is the basis of the technique; each scattered electron carries energetic information, which can be converted in a spectrometer into a loss spectrum. The spatial resolution in EDS and EELS is defined laterally by the diameter of the incident beam (< 1 nm for the latest generation of instruments), and vertically by the dimension of the specimen in the direction of the incident beam (if the specimen is sufficiently thin so that diffuse scattering can be neglected). Thus, both techniques are appropriate for analytical interrogation on the nanoscale.

A schematic is shown in Figure 3.1 of the various scattering processes that may take place during interaction between incident energetic electrons and a thin specimen. This chapter will focus on the resultant generation of characteristic X-rays, and on inelastic losses from the primary beam.

3.2 LOSS PROCESSES

The functional dependence of the intensity of the X-ray continuum, I_{cm}, (also known as 'Bremsstrahlung') is given by

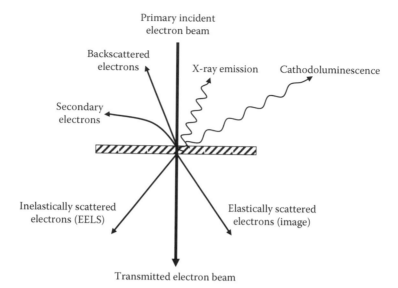

FIGURE 3.1   The possible scattering processes arising from interaction between an energetic electron beam and a thin specimen. Emission of characteristic X-rays, and inelastic scattering, are the two most relevant processes leading to compositional information (by EDS) and chemical information (by EELS). Cathodoluminescence arises from promotion of electrons across the bandgap of a semiconductor or an insulator, due to the impact of energetic electrons, or ions, followed by de-exitation by emission of one or more photons. It is widely used by the semiconductor industry and for investigations of ceramics and minerals. In the context of nanostructural characterization, it is not a widely used technique.

$$I_{cm} \propto i_p Z(E_0 - E_v)/E_v \qquad (3.1)$$

where i_p is the incident beam current, Z is the average atomic number, E_0 is the incident beam energy, and E_v is the continuum photon energy at some point in the spectrum. The intensity decreases as the continuum energy increases, and also decreases with decreasing atomic number.

Characteristic X-ray lines will be superimposed on the continuum spectrum. The mechanisms that account for these lines are shown schematically in Figure 3.2.

3.2.1   EDS Spectral Notation

The X-ray 'shell' notation is used to designate the electronic states that play a role in the EDS process. The shell closest to the nucleus is designated K, and is occupied by two electrons (in an 1s-orbital); the next shell is the L, with an occupation of eight electrons (two in a 2s-orbital plus six in 2p-orbitals), followed by the M shell with an occupation of eighteen electrons (two in 3s-orbitals, plus six in 3p-orbitals, plus ten in 3d-orbitals), and so on. The spectral resolution of an EDS detector is in the range 50–150 eV. Accordingly, the fine structure splittings that account for sub-levels within a shell (e.g., spin-orbit splitting), beyond those that account for the sub-levels

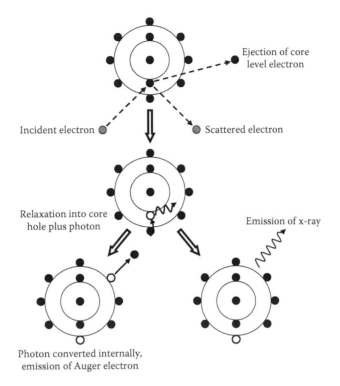

FIGURE 3.2 Schematic of the processes that account for the characteristic lines appearing in an EDS spectrum. The initial event is that of ejection of a core electron, thus leaving the atom in an excited state. In the initial relaxation route, an electron from a higher energy level drops into the original core hole; additionally, there is a transient generation of a local photon. Following that, there are two possible relaxation channels, one of which will lead to emission of an Auger electron (being the object of interest in Auger electron spectroscopy), while the other leads to emission of an X-ray quantum with characteristic energy, which is the object of interest in EDS.

corresponding to the total angular momentum quantum number, cannot be resolved, and the X-ray shell notation, K, $L_{I,II,III}$, $M_{I,II,III,IV,V}$, and so forth, is a sufficient label for an X-ray peak in the EDS spectrum. The groupings of the X-ray families, and their notation, are shown schematically in Figure 3.3.

3.2.2 EDS Spectra

The dependence of the X-ray energies of transitions as a function of atomic number is shown in Figure 3.4. For practical reasons, discussed below, only the energy range from a few hundred eV up to ca. 20 keV is available for analytical work. It is also obvious that the complexity of the resultant spectrum increases with atomic number.

The energy of the characteristic X-ray line, E, within a given series of lines, increases monotonically with atomic number. The dependence is known as 'Moseley's Law',

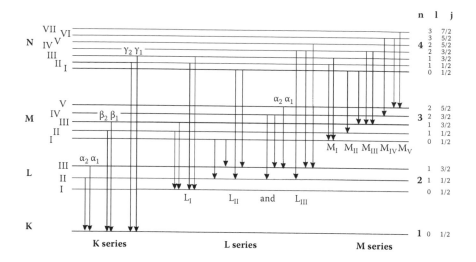

FIGURE 3.3 The X-ray families are defined by the principal quantum number n = 1, 2, 3, etc., (K, L, M, N in X-ray notation). There are sub-shells defined by the total angular momentum quantum number j = ℓ ± ½, where ½ comes from the electron spin quantum number s. Accordingly, the K-shell has no sub-shell, while L has three, M has 5, etc. Allowed transitions are determined by the selection rule Δℓ = ± 1. Thus, there are six allowed transitions from the L, M, and N shells to the K shell, fourteen from the M and N shell to the L sub-shells, nine from the N shell to the M sub-shells, and so on. The Greek letter notation, α_1, α_2, β_1, β_2, etc., refers to the level from which the transition originates.

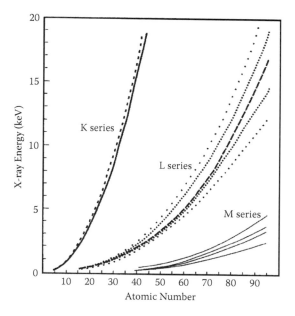

FIGURE 3.4 The dependence of the energy of characteristic X-ray transitions on atomic number for the K, L, and M series.

$$E = C_1(Z - C_2)^2 \qquad (3.2)$$

where Z is the atomic number of the emitting atom, and C_1 and C_2 are constants. Moseley's Law is the basis for 'finger-printing' elemental analysis by EDS. If the energy of a given K, L, or M line is measured, then the atomic number of the element producing that line can be identified (Goldstein et al. 1992).

3.2.2.1 Fluorescence Yield (ω)

The cross-section for ionization leading to the emission of a particular X-ray is expressed by the fluorescence yield, ω, as the ratio of the number of X-ray photons produced divided by the number of shell ionisations, where the sum of the fluorescence yield and Auger yield is unity. The dependence on atomic number and shell from which the primary electron is ejected from a core state is shown in Figure 3.5.

Within a given series of lines, ω increases with atomic number and, for a given atomic number, is greatest for K shells and progressively less for L and M shells. The obvious implication is that the sensitivity factors for high-Z elements are lower than those for low-Z elements.

3.3 EELS

An EELS spectrum is a far richer source of information than that of EDS, but compositional quantification by EELS is more difficult and less reliable than that for EDS. The information content of EELS is illustrated in Figure 3.6.

3.3.1 EELS Spectral Features

The following structures will/may be present in order of increasing energy loss:

Zero-Loss Peak: A dominant zero-loss peak, which accounts for electrons that have lost no energy on their way through the specimen (except for small

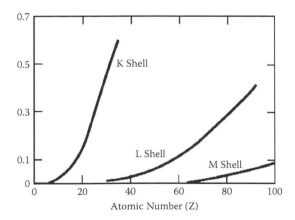

FIGURE 3.5 Fluorescence yield (vertical axis) as a function of atomic number and shell.

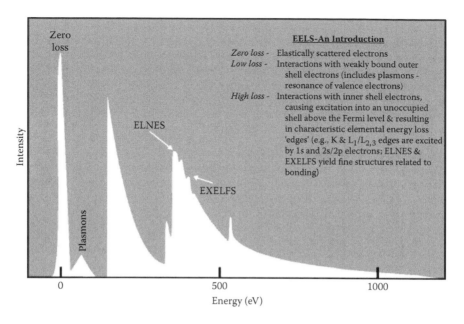

FIGURE 3.6 Information content of an EELS spectrum. The loss spectrum beyond the plasmon structure has been expanded by several orders of magnitude.

losses owing to phonon scattering). The energy width of the zero-loss peak is a measure of the spread in energy of the electron source.

Plasmon Structure: Plasmon structure is due to interactions between incident electrons and quasi-free electrons in the valence and conduction band/states, leading to collective excitations and longitudinal charge-density waves. In thick samples, multiple sequential electron-plasmon excitations can occur, and will appear in the spectrum as evenly spaced peaks of decreasing amplitude. As well, ionisation of electrons from bulk states close to the Fermi energy may contribute. Excitation of surface plasmons can also play a role, giving rise to transverse charge-density waves.

Ejection of Electrons from Core States: The intrinsic linewidth of core-state energies is less than 1 meV. However, a core-state energy can shift by up to several eV in response to the chemical environment of the atom; this is known as the 'chemical shift', and is a major focus of interest in X-ray photoelectron spectroscopy (XPS) (Briggs and Seah 1990). While the presence of ionization edges constitutes finger-print information about the presence of atomic species, the intensities of the edges provide semi-quantitative compositional information, and the chemical shifts provide information about the chemical environment (e.g., bonding). The great merit of EELS is therefore that compositional and chemical information can be obtained with near-atomic lateral spatial resolution. As distinct from EDS, the spectral features that are associated with core-state ionisation are not dependent on the fluorescence yield, and are suitable for the analysis of low-Z elements. The minimum energy, E_c, for ionisation, (the binding energy of the inner-shell electron to the nucleus), is the location of the edge. This is also where the ionisation cross-section reaches its maximum. However, ionisation will also occur with larger energy losses, when $E > E_c$, but the cross-section decreases with increasing energy loss, due to multiple low-loss interactions. Thus, the intensity of the absorption edge decreases with increasing energy loss, following a power-law dependence.

When single nanostructures are analysed by EELS, it is the characteristic ionisation edges that are of primary interest, followed by the respective chemical shifts. In general, the count rates will be too low to obtain much useful information from the subtler features in the spectra.

ELNES Structures (E_c to E_c+50 eV): In close proximity to the ionization edges there will be energy-loss-near-edge-spectroscopy structures (ELNES). The near-edge structure arises from coupling of the scattered electron with core states, with electrons being promoted into unoccupied conduction band states. The spectrum arises from the relative probabilities of particular transitions taking place.

EXELFS Structures (beyond E_c+50 eV): Extended-electron-energy-loss fine structure (EXELFS) is also observed in the vicinity of the ionization edge, and will sometimes overlap with the ELNES structure. It is the electron-excited analogue of the X-ray-induced extended X-ray absorption fine structure (EXAFS) technique. The ejected electron interacts with the local short-range order of the atomic environment; the result is a periodic oscillation in intensity due to constructive and destructive interactions between electron waves stimulated by the energetic electron initially ejected from a core level. The modulations contain information about interatomic distances and coordination.

The quality of the information from loss spectroscopy is critically dependent on the linewidth of the primary incident beam. The latest advances in energy-filtered (S)TEM instruments have resulted in linewidths of ca. 0.15 eV at full width half-maximum (FWHM). This is rather better than the linewidths available in the monochromatic X-ray sources used for XPS analysis. Accordingly, chemical shifts can be determined by EELS with greater spectral resolution than in XPS. While plasmon structures and ionization edges are relatively prominent features in an EELS spectrum, ELNES and EXELFS spectra are much weaker, and require either high luminosity and/or long counting times. Thus, there are problems arising from instrumental drift and radiation damage.

3.4 TECHNICAL IMPLEMENTATION AND METHODS

3.4.1 EDS

The critical part of an EDS spectrometer is the energy-dispersive detector, shown in Figure 3.7. The aperture/collimator (1) defines the solid angle, $\Omega/4\pi$, typically 0.3 steradian, through which incoming X-rays are collected; the collimator also helps to exclude stray signals from components in the vicinity of the sample. The electron trap (2) sweeps out stray electrons that enter the aperture, and prevents them entering

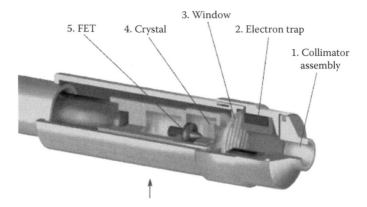

FIGURE 3.7 **(See color insert.)** The energy-dispersive EDS detector. (From http://www. charfac.umn.edu/instruments/eds_on_sem_primer.pdf.)

the active part of the detector. The thin window (3) isolates the vacuum in the detector region from that of the instrument, thus protecting the active part of the detector from contamination. Instruments with field-emission sources are generally operated under clean, near-ultra-high vacuum (UHV), conditions. Window-less detectors can then be used, leading to enhanced detector sensitivity for low-energy X-rays, as shown in Figures 3.8 and 3.9 (Schlossmacher et al. 2010). The enhancement at low X-ray energies is particularly relevant for K-shell transitions and low-Z elements that are prominent constituents of nanostructures (e.g., graphene, carbon nanotube (CNT), boron nitride (BN) nanotubes, etc.). The crystal (4), generally single crystal Li-drifted Si, is the critical energy dispersive element. Incoming X-rays of energies

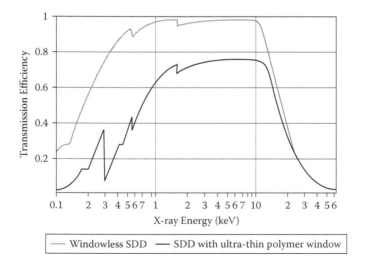

FIGURE 3.8 Plot of transmission efficiency versus X-ray energy, showing that window-less EDS silicon drift detectors (SDD) have enhanced efficiency for low-energy X-rays. (From Schlossmacher et al., 2010, *Microscopy and Analysis*, November, p. 55.)

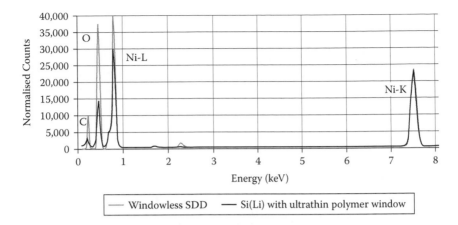

FIGURE 3.9   EDS spectra for an identical specimen obtained under identical conditions, apart from the change in detector, illustrating enhanced count rates at low energy. (From Schlossmacher et al., 2010, *Microscopy and Analysis*, November, p. 55. With permission.)

up to ca. 10 keV undergo total absorption in the crystal, by a succession of photo-electron ejection processes. The photo-electrons give rise to additional ionization events which themselves cause generation of electron-hole (e-h) pairs. On average one e-h pair is produced for a loss of 3.8 eV. Thus, a 1 keV X-ray will generate ca. 300 e-h pairs. An axial **E**-field of about 2×10^5 V/m is applied across the crystal and sweeps electrons and holes to their respective electrodes, thereby creating a current pulse. The current pulse is sensed by a high input impedance field effect transistor (FET) I/V converter (5). From there additional amplification, pulse-shaping, and filtering is carried out before the pulse is passed to a multi-channel analyzer. The important point to note is that the amplitude of the pulse is directly proportional to the energy of the incident X-ray. The spectral FWHM resolution typically ranges from ca. 50 to ca. 150 eV for sub-keV to 5 keV X-rays, respectively. Due to the relatively low hole mobility, the slow rise time of the current pulse limits the count rate in EDS (generally less than 10^5 s^{-1}). The implication is that the detector has a relatively long 'dead' time. If a second X-ray arrives during the dead time it will not be counted, leading to an erroneous outcome. A tolerance of 25% is usually an acceptable compromise between counting time and missed counts.

The length of the active part of the crystal is also a compromise between the requirement for high count rate (shorter length) and the requirement for total absorption of high-energy X-rays (greater length). Usually the length is chosen so that X-rays with energies up 10 keV are absorbed; characteristic X-rays of energies below 10 keV are emitted from all elements (see Figure 3.7).

3.4.1.1   Quantification of EDS Spectra

Elemental quantification by EDS has been described in detail in several monographs (Russ 1984; Williams, Goldstein, and Newbury 1995; Titchmarsh 2009). Quantification generally proceeds from the premise that the specimen is of semi-infinite size, and is homogeneous and amorphous. In the case of nanostructures,

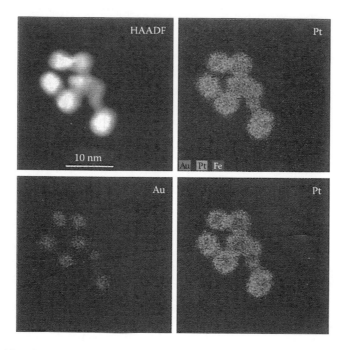

FIGURE 3.10   (**See color insert.**) Compositional mapping by EDS of particles in the 5 nm size range. (From Schlossmacher et al., 2010, *Microscopy and Analysis*, November, p. 55. With permission.)

none of those assumptions is valid. Accordingly, EDS is generally used as a qualitative finger-printing tool for compositional mapping and contour line analysis, see Chapters 8 and 9, rather than as a means of obtaining reliably quantifiable composition. Furthermore, the best results are obtained with the use of standards; in the case of SEM/EDS and WDS a range of standards can be located permanently and accessibly within the sample chamber. On the other hand, the use of standards for TEM/EDS and STEM/EDS becomes impracticable due to the frequent need to load and unload series of specimens. In this case, resort has to be made to the so-called 'standard-less' quantification procedure. In general, EDS systems come with the manufacturer's software which includes step-by-step routines for quantification.

An example (see Figure 3.10), of compositional mapping by EDS of particles in the 5 nm size range, illustrates the current state-of-the-art.

3.4.2   EELS

The EELS spectrometer is essentially a magnetic sector band-pass filter, as shown in Figure 3.11. The most widely used arrangement, developed by Gatan, Inc. (Krivanec et al. 1992; Gubbens et al. 1998), disperses the beam after its arrival in the image plane, in TEM, or after it has passed through the specimen, in STEM. The angle selection slit limits the scattering angle of electrons emerging from the image plane (or sample). The magnetic sector has the effect of focussing electrons in the plane in which the energy-selecting slit is located. The width of that slit determines the width

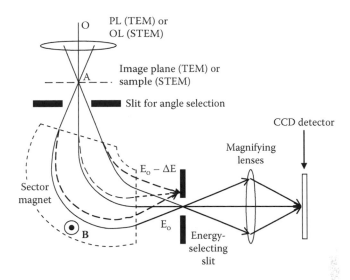

FIGURE 3.11 Schematic of arrangement for EELS analysis. (From Titchmarsh, 2009, with permission from Taylor & Francis.)

of the pass-band; electrons with energies outside the pass-band will be rejected. Additional optics focus the divergent beam emerging from the selection slit onto a CCD detector plate. For EELS in combination with TEM, the spectrometer will produce an energy-filtered image by parallel data acquisition, while for STEM/EELS an energy-dispersed spectrum is obtained at each location in the specimen by serial data acquisition.

The spatial resolution for the combination TEM/EELS depends on the operational mode, that is, imaging or diffraction. In the former case, the interaction volume of the signal corresponds to the demagnified image of the spectrometer aperture in the object plane. For a magnification of 10^5, and an aperture diameter of 1 mm for the spectrometer, the spectrum will arise from an area of diameter of 10 nm. The effect of chromatic aberration may degrade this by a factor of ten. Thus, the energy resolution may be good, but the spatial resolution will be modest. In the diffraction mode, however, the spatial resolution will approach that of the beam diameter at the specimen. In a STEM instrument, there are no optics beyond the specimen, and the spatial resolution can under favourable circumstances be that of a single column of atoms.

The spectral resolution, ΔE, can be no better than the FWHM of the energy spread of the primary beam, that is, the FWHM of the zero-loss peak. For a cold cathode source, ΔE is typically 0.3 eV, which can be improved by a factor of two by introducing a monochromator into the source assembly (Batson 1999; Freitag et al. 2005).

3.5 COMPLEMENTARITY OF EDS AND EELS: A CASE STUDY

The respective merits of EDS and EELS in combination with S/TEM can best be illustrated by a case study. A well-designed and executed example can be found on the Gatan Inc. Web site (Kundmann 2007). The following discussion is based, with

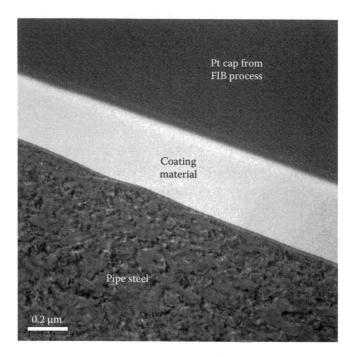

FIGURE 3.12 TEM survey cross-section image of a coating, sandwiched between a steel substrate and a Pt cap introduced as part of the FIB milling process. (From Kundmann, 2007, http://www.gatan.com/resources/answers/1.php.)

some adaptations, on material from that Web site. While the example does not relate specifically to nanostructures, much of the methodology is relevant (e.g., for nanotubes, quantum dots, etc.). Other examples that are more directly concerned with nanostructures can be found in the chapters in Section II of the book.

The sample to be investigated consisted of a multilayer wear- and corrosion-resistant coating on a steel pipe (developed by D. Upadhyaya of Sub-One Technology). A cross-section thin foil specimen containing the layer structure and substrate was prepared by focussed ion beam milling.

A zero-loss filtered bright field TEM image (see Figure 3.12) shows an overview of the interface region. A more detailed image, obtained in the STEM mode, of the coating and an interfacial layer is shown in Figure 3.13.

In general, the best analytical results will be obtained when the thickness of a foil is comparable to the mean free path for inelastic scattering of the incident beam. The reason is that multiple inelastic collision events lead to loss of information, and thus to a lower signal-to-noise ratio. When the thickness of the foil is comparable to the inelastic mean free path, then the effect of multiple events is minimised. The line profile, obtained by a linear raster of the beam in the STEM HA-ADF mode, for the zero-loss line, in Figure 3.14, shows that the specimen in the region of interest (ROI) met that requirement. The profile is essentially derived from Z-contrast, where a low/high average atomic number gives rise to lower/greater angular scattering, and thus lower/greater intensity on the ADF detector. Accordingly, it is apparent that the

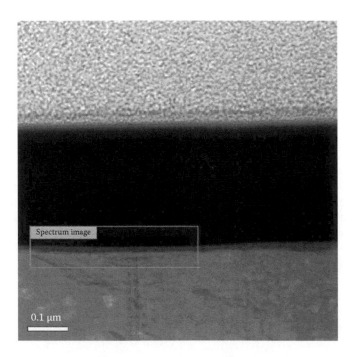

FIGURE 3.13 STEM image of the coating and the adjacent interfaces. An interfacial layer between the coating and the steel substrate is visible. The rectangular box shows the region from which analytical information was obtained. (From Kundmann, 2007, http://www.gatan.com/resources/answers/1.php.)

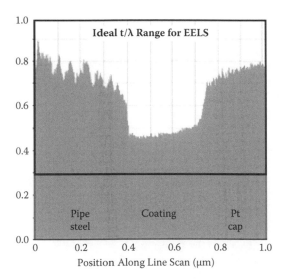

FIGURE 3.14 Thickness profile along a line perpendicular to the interface coating. The vertical scale is in units of mean free path, λ, for inelastic scattering. (Adapted from Kundmann, 2007, http://www.gatan.com/resources/answers/1.php.)

average atomic number in the region of the coating was lower than that in the substrate and in the Pt cap.

The rectangular region, shown in Figure 3.13, was analyzed by EELS and EDS recorded in parallel at 160×40 points spaced 2.7 nm apart, producing a data set containing 6400 pairs of spectra. The total data acquisition time was ca. 6 min. EELS data were acquired over the energy-loss range from 50 to 1050 eV with a collection angle of ca. 17 mR and a convergence angle of 10 mR, while EDS data were acquired for X-ray energies from 0.2 to 20 keV.

There are several advantages to the collection of a large database of simultaneous pairs of spectra over a region of interest, as opposed to point or line analyses at selected locations. For instance, spurious results due to artefacts can be ruled out immediately, and there is a permanent record of spatially resolved spectral information that can be analysed at leisure, so that the results may be examined in greater detail to reveal anything unexpected that was not noticed during the measurement session. Finally, any effects due to specimen damage, (e.g., in carbon-based structures, see Chapter 9), accumulation of carbon contamination, and/or mechanical or electronic drift can be assessed. As a result, it is possible to improve the signal-to-noise ratio by summing selected spectra.

The energy-filtered TEM (EFTEM) and Z-contrast images of the 20 nm thick layer between the steel and the coating clearly suggest that the layer has a composition with a relatively high oxygen content, as might be expected for a surface oxide layer on the steel.

Four representative regions of interest (see Figure 3.15) were identified and selected for quantitative EELS and EDS analysis.

Regions 1 and 4 are representative of the bulk of the Pt coating material and of the steel substrate, respectively. Region 3 is centred on the 20 nm interface layer associated with the steel substrate, while region 2 corresponds to the very thin layer at the junction between the 20 nm layer and the Pt coating. Cumulative EELS and EDS spectra extracted from these regions are shown below (see Figure 3.16), along with approximate atomic ratios extracted quantitatively from the spectra.

The spectral data focus on signals from the low-Z elements, that is, C, O, Si, and also Ar, which was implanted during the focussed ion beam machining of the specimen. The spectra illustrate the method whereby semi-quantitative compositional data

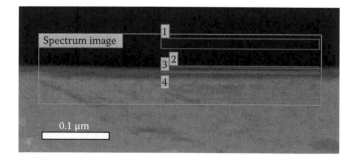

FIGURE 3.15 (**See color insert.**) The four regions of interest selected for EELS and EDS analyses are shown. (From Kundmann, 2007, http://www.gatan.com/resources/answers/1.php.)

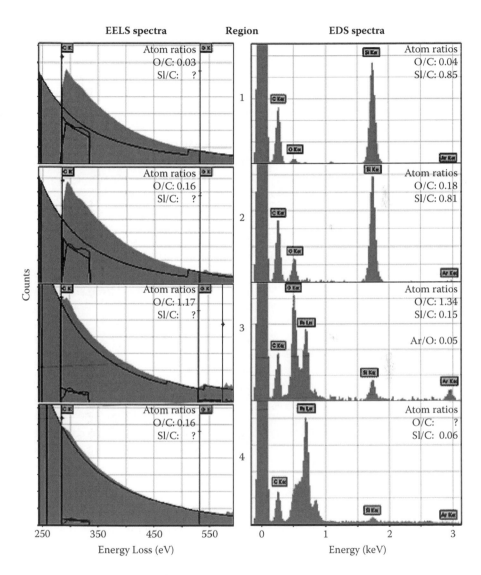

FIGURE 3.16 Cumulative EELS and EDS spectra for the four regions of interest shown in Figure 3.15. (Adapted from Kundmann, 2007, http://www.gatan.com/resources/answers/1.php.)

can be extracted from EELS spectra. A polynomial curve is fitted to the background above an ionisation edge, so that the height of the edge is a measure of the compositional abundance of the species corresponding to the edge feature. When empirical sensitivity factors are applied, ratios of abundances can be calculated. It is noteworthy that the agreement with the more reliable EDS compositional data is good. This may not always be so, however, and EELS is at best a semi-quantitative technique.

The data in Figure 3.17 are concerned with the spectral features of the transition metal and oxygen features. Although these are apparently two separate views of the data, it should be kept in mind that the spectra were recorded simultaneously during data acquisition.

The spectra in Figures 3.16 and 3.17 demonstrate the value of the complementarity of EELS and EDS analyses. For instance, identification and quantification of the Si L-edge is difficult from the EELS spectra, but is relatively straightforward and more reliable by EDS. Conversely, quantification of oxygen by EDS in the presence of the transition metals, Ti, V, Cr, and Mn is difficult but the relevant features are well resolved in the EELS spectra and can be used for semi-quantitative analyses. The Cu peaks in the EDS spectra are spurious and originate from the support grid and/or components in the TEM environment, and represent a commonly encountered problem. However, EELS measures primary scattering events of the direct beam; accordingly, any spurious signals would be below the threshold of detection.

The most significant finding, based on the analytical data, is that region 2 appears to be predominantly an oxide of silicon, that is, it does not involve any of the metal species from the steel substrate. The spectra obtained from region 3 represent the native oxide layer on the steel substrate.

The conclusions are supported by the details of the EELS data in the vicinity of the oxygen K-edge, by considering the energy-loss near edge structure (ELNES). The ELNES structures reflect the unoccupied density of states in the vicinity of the Fermi level, and will be dependent on the formal valence and bonding.

In Figure 3.18 are shown the O K ELNES spectral for the two distinct interfacial oxide layers proposed for regions 2 and 3.

The presence and absence, respectively, of the shoulder in the O K-edge spectra measured from Fe_2O_3 and SiO_2 reference samples (see Figure 3.19) correlate well with the hypothesised oxide regions.

Information obtained from EELS analysis is particularly valuable for low-Z nanostructures. For instance, BN nanotube spectra exhibit a prominent B K-edge at ca. 200 eV, while the N K-edge is a less prominent, but readily observable, structure at ca. 400 eV. By comparison, the corresponding EDS transitions are only marginally detectable due to low fluorescence yield and poor detector efficiency. As well, oxygen is a ubiquitous contaminant in a TEM environment.

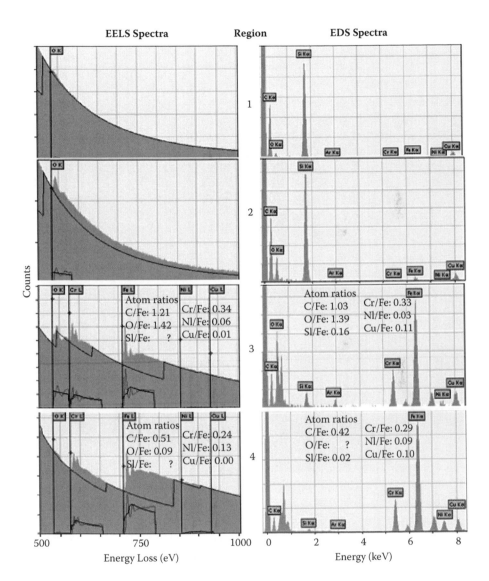

FIGURE 3.17   EELS and EDS spectral information for O and transition metal constituents. (Adapted from Kundmann, 2007, http://www.gatan.com/resources/answers/1.php.)

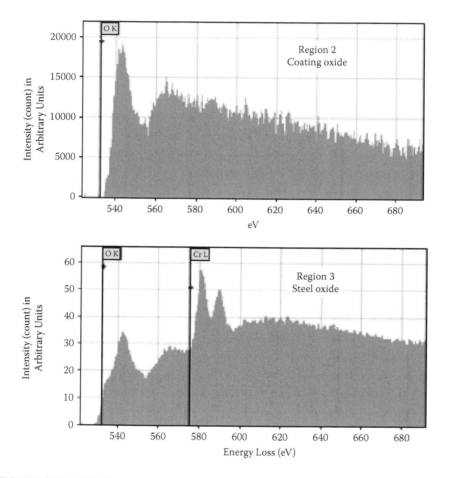

FIGURE 3.18 EELS spectral features in the vicinity of the O K-edge for the two oxide regions. (Adapted from Kundmann, 2007, http://www.gatan.com/resources/answers/1.php.)

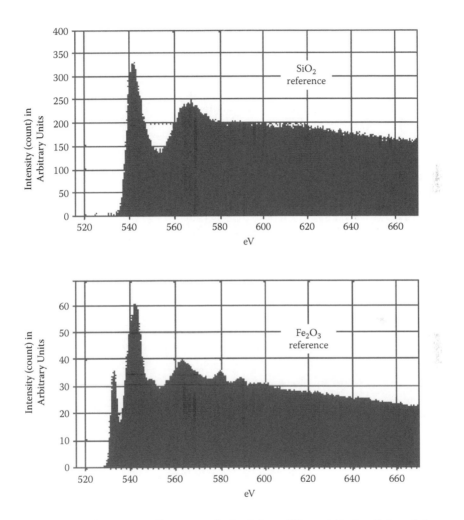

FIGURE 3.19 Reference EELS spectra obtained from SiO_2 and Fe_2O_3. (Adapted from Kundmann, 2007, http://www.gatan.com/resources/answers/1.php.)

ACKNOWLEDGEMENTS

We wish to acknowledge valuable discussions with Dr. K. Jurkschat, and with Dr. R. Nicholls.

REFERENCES

Ahn, C. C., (ed.), 2004, *Transmission Electron Energy Loss Spectrometry in Materials Science and the EELS Atlas,* 2nd ed., Darmstadt: Wiley-VCH.

Batson, P. E., 1999, *Ultramicroscopy*, 78, 33.

Briggs, D. and Seah, M. P. (eds.), 1990, *Practical Surface Analysis*, 2nd ed., New York: Wiley.

Brydson, R., 2001, *Electron Energy Loss Spectroscopy*, Boca Raton: Garland Science.

Egerton, R. F., 2009, *Rep. Prog. Phys.* 72, 016502.

Freitag, B., Kujawa, S., Mul, P. M., Ringnalda, J., and Tiemeijer, P. C., 2005, *Ultramicroscopy*, 102, 209.

Goldstein, J. I., Newbury, D. E., Echlin, P., Joy, D. C., Romig, A. D, Lyman C. E., Fiori, C. E., and Lifshin, E., 1992, *SEM and X-Ray Microanalysis*, London: Plenum Press.

Gubbens, A. J., Brink, H. A., Kundman, K. A., Friedman, S. L., and Krivanek, O. L., 1998, *Micron*, 29, 81.

Krivanek, O. L., Gubbens, A. J., Delby, N., and Meyer, C. E., 1992, *Microsc. Microan. Microstr.*, 3, 187.

Kundmann, M., 2007, *Need-to-know Information on Microscopy Applications*, Issue 16, July, http://www.gatan.com/resources/answers/1.php.

Russ, J. C., 1984. *Fundamentals of Energy Dispersive X-Ray Analysis*, London: Butterworths.

Schlossmacher, P, Klenov, D. O., Freitag, B., von Harrach, S., and Steinbach, A., 2010, *Microscopy and Analysis*, November, p. 55.

Titchmarsh, J., 2009, In *Handbook of Surface and Interface Analysis*, 2nd ed., Rivière, J. C. and Myhra, S., (eds.), Boca Raton, FL: CRC.

Williams, D. B., Goldstein, J. I., and Newbury, J. E. (eds.), 1995, *X-Ray Spectrometry in Electron Beam Instruments*, London: Plenum Press.

4 Photon-Optical Spectroscopy—Raman and Fluorescence

4.1 INTRODUCTION

Raman and fluorescence spectroscopies are among the most widely used techniques for the characterization of nanostructures. At first glance, this may seem an unlikely situation, given that the spatial resolution of both techniques is diffraction limited, and can be no better than 0.25 µm for their confocal probe versions. Raman spectroscopy is favoured for carbon-based structures, where it couples with characteristic intramolecular vibrational modes or with lattice vibrational modes for those structures with crystalline order. Fluorescence spectroscopy plays an important role for quantum-well structures where it couples with the atom-like electron states. Both techniques require minimal sample preparation, and deliver unequivocal finger-print spectra from which qualitative and semi-quantitative information can be inferred.

4.2 RAMAN SPECTROSCOPY

4.2.1 PHYSICAL PRINCIPLES

When a molecule, or a material with long-range order, is illuminated with light, there will be scattering of the incident photons. If the scattered photon emerges from the interaction with its original frequency, the process is elastic and is known as 'Rayleigh scattering' (sometimes the Mie theory has to be considered in order to describe the result, see Chapter 6). Inelastic scattering as a result of energy loss can also occur, and is then known as 'Raman scattering'. The consequence of the latter is a change in frequency, or a shift in wavelength of the photon, to either higher or lower (Stokes and anti-Stokes, respectively) frequency (i.e., the photon loses or gains energy). The loss, or gain, is discrete due to coupling of the photon with the vibrational or electronic states of the molecule or crystalline material. Since these states are specific to the intramolecular structure, or to the crystal structure, the resultant spectral features will be characteristic of the material being illuminated. In the present context, of nanostructural characterization, the distinction between molecular structure, on the one hand, and crystalline materials with long-range order, on the other hand, will not be clear-cut. While the underlying phenomena are the same, the theories tend to be distinctly different. A full and complete treatment of the theory of Raman scattering is beyond the scope of this volume (the reader is referred to the specialist literature, e.g., Nakamoto 2009a; Nakamoto 2009b; Turrell and Corset

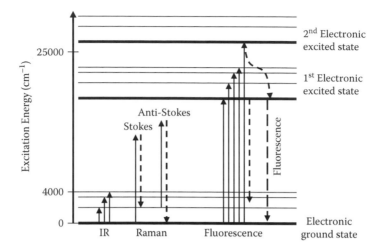

FIGURE 4.1 Schematic summary of the types of excitation and relaxation processes that are accessible with IR, Raman, and fluorescence spectroscopy. The photon energies are given in units of wave number. The theory is usually framed in terms of frequency and frequency shifts. Experimental results are generally quoted in terms of wave numbers and wave-number shifts, $\bar{\nu}$, in units of cm^{-1}. The conversion is $\bar{\nu} = \nu_{vib}/c$, where c is the speed of light. The conversion to energy is, therefore, that 1 eV is equivalent to 8.0655×10^3 cm^{-1}. (Adapted from Schwartz, 2011.)

1996; Lin-Vien, et al. 1991; Turrell 1972). Instead, a basic classical picture will be developed in order to gain some insight into the process.

The schematic in Figure 4.1 gives an overview of the scattering processes that are relevant to optical vibrational spectroscopy. The 'classical' Raman effects include the Stokes and anti-Stokes effects, where the initial excitation takes the system into a virtual state.

The summary in Figure 4.1 illustrates some important aspects of photon spectroscopy.

- Infrared (IR) IR spectroscopy is based on the spectral dispersal of low-energy radiation that has undergone inelastic loss due to coupling with intra-molecular vibrational modes. The analysis can be carried out in either the transmission/absorption or reflection modes. In some important respects, it can be considered complementary to Raman spectroscopy, in that the latter couples with symmetric (*gerade*) vibrational modes, while IR couples with anti-symmetric (*ungerade*) vibrational modes.
- The differences in the relative intensities of the Stokes and anti-Stokes lines are due to the population in the ground state being greater than that in the first excited state. Accordingly, the Stokes transitions are the ones that are usually investigated.
- Resonance Raman analysis requires a tunable laser source and an extended range spectrometer with which the electronic excited states can be probed.
- Fluorescence can present a problem in Raman spectroscopy. Many organic and biomaterials fluoresce in the visible spectral range at an intensity that may swamp completely the Raman signals. One remedy is to use a longer

wavelength excitation source, and thus suppress the fluorescence yield. The other is to 'bleach' the material with a lengthy period of irradiation before commencing Raman analysis. On the other hand, fluorescence is what is analysed in fluorescence spectroscopy (see Section 4.3 in this chapter). The significant difference is that in fluorescence the incident photon is completely absorbed by the system, which is then transferred to an excited state. The excited state will then relax back to its ground state usually via a number of intermediate states, where at least one transition gives rise to the emission of a characteristic photon. Fluorescence can therefore be considered to be a resonance process.

4.2.2 A FORMAL CLASSICAL DESCRIPTION OF THE RAMAN PROCESS

Lagrangian formalism can be deployed to transform a system with a number of vibrational and rotational degrees of freedom into a simpler form described by generalized coordinates (known as 'normal coordinates, Q', where Q can be a spatial or angular variable). The merit of the transformation is that each vibrational mode can be described by a single independent equation (e.g., Landau and Lifshitz 1960).

If a molecule is vibrating or rotating in a sinusoidal fashion, the change in the normal coordinate ΔQ can be written as a sinusoidal function of the frequency of the vibration, ν, and the time, t. viz.,

$$\Delta Q = Q_{max} \cos(2\pi\nu t) \tag{4.1}$$

where Q_{max} is the maximum vibrational amplitude. Incident light that has a particular frequency, ν_0, will induce an electric field, E, that also has sinusoidal behaviour, given by

$$E = E_0 \cos(2\pi\nu_0 t) \tag{4.2}$$

Here, E_0 is the amplitude of the electric field, and ν_0 is the frequency.

In general, the response of a polarizable material to incident electromagnetic radiation can be described by a matrix equation of the form

$$\bar{P} = [\bar{\alpha}_{ij}] \times \bar{E}_0 \cos(2\pi\nu_0 t) \tag{4.3}$$

where $\bar{P}$ is the polarization vector, and $\bar{\alpha}_{ij}$ are the elements of a 3×3 polarization tensor.

The tensor elements are non-linear functions of the applied field, and it is common practice to expand them in a Taylor series which is then terminated at the leading first-order term. The expansion is carried out in terms of the normal coordinates, Q, of the vibrational modes, which describe the dynamics of the system.

$$\alpha_{ij} = \alpha_{ij}^0 + \left(\frac{\partial \alpha_{ij}}{\partial Q}\right)_{Q=Q_0} \times Q \qquad (4.4)$$

where i and j range over the Cartesian coordinates. The response of the material in a direction corresponding to the polarisation vector of the E-field can now be described by

$$P_i = \sum_j \alpha_{ij} E_j = \sum_j [\alpha_{ij}^0 E_{0j} E_{0j} \cos(2\pi \nu_0 t)]$$

$$+ \frac{E_{0j} Q_0}{2}\left(\frac{\partial \alpha_{ij}}{\partial Q}\right)_{Q=Q_0} [\cos(2\pi(\nu_0 - \nu_{vib})t) + \cos(2\pi(\nu_0 + \nu_{vib})t)] + h.o.t. \qquad (4.5)$$

The zeroth-order term represents Rayleigh scattering, which is elastic, while the two first-order terms, with the factors $(\nu_0 - \nu_{vib})$ and $(\nu_0 + \nu_{vib})$, describe Stokes and anti-Stokes scattering, respectively, which correspond to the Raman effect. It should be noted that the Raman terms are non-zero only if

$$\left(\frac{\partial \alpha_{ij}}{\partial Q}\right)_{Q=Q_0}$$

is non-zero (Lewis and Edwards 2001). There may also be contributions from higher-order terms (h.o.t.).

From Equation (4.5), it is apparent that Raman spectroscopy depends on both electrical (α_{ij}) and mechanical (ν_{vib}) properties of the materials under investigation. Accordingly, the spectra will depend on, and reveal information about:

(i) The attributes that determine the mechanical vibrational frequencies, such as atomic mass, and interatomic force constants, and/or the geometry (inter-atomic distances, atomic substitutions). These attributes will determine the spectral locations of the modes.

(ii) The attributes associated with the polarizability (such as ionicity/covalency, band structure, electronic insertion), which will determine the extent of vibration-induced variation in charge that is taking place, and thus deter-mine the intensity of a mode.

4.2.3 SERS—SURFACE ENHANCED RAMAN SPECTROSCOPY

Conventional Raman spectroscopy is based on an inefficient process, because only one in every 10^7 incident photons undergoes a Raman shift. It has been discovered that if a molecule under investigation were to be adsorbed onto a nanostructured noble metal substrate, preferably Ag or Au, then the efficiency of the process could be increased by many orders of magnitude. This phenomenon has become known as 'surface enhanced Raman spectroscopy' (SERS).

The mechanism is thought to be that of excitation of surface plasmons in the metallic substrate by the incident radiation, due to the matching of the frequency of the incident radiation with that of the plasmon. The result is the stimulation of strong near-field radiation in which the adsorbed molecular species, or nanostructure, will be submerged. However, there is evidence for additional effects related to specific interactions between the adsorbate and the nanostructure of the substrate metal (see Kneipp et al. 2002; Moskovits et al. 2002; Otto 2002).

The SERS phenomenon, in combination with the latest generation of probe instruments with parallel detection, is particularly relevant to nanostructures since it has been shown that spectra can be obtained from single molecules, or single nanostructures (e.g., Nie and Emory 1997).

4.2.4   TERS—Tip Enhanced Raman Spectroscopy

While SERS can obtain a Raman signal from a single molecule, the localization in the x-y plane of the molecule is limited by the optical spatial resolution (no better than $\lambda/4$ in the case of a confocal instrument). Scanning near-field optical microscopy (SNOM, also known as 'NSOM') is an optical technique based on the technology of the earlier scanning probe instruments (STM and AFM, see Chapter 5). By exploiting the spatial decay of the optical near-field it has been possible to map structures optically in transmission or in backscatter modes with a lateral resolution down to ca. 30 nm. The conventional way of confining the near-field radiation laterally was to use a fibre-optic tip terminated by an aperture of the same lateral dimension as the desired resolution. Due to the resultant limitation of the flux of incident photons, implementation of optical spectroscopy at that level of spatial resolution has proved to be a challenge. The more efficient fluorescence process can produce useful results with the conventional probe arrangement, but the less efficient Raman process will not result in usable information.

The latest generation of SNOM instruments has now dispensed with the fibre-optic probe in favour of an illuminated sharp metallic tip. The illumination generates the near-field illumination of the specimen by stimulation of surface plasmons in the tip. The arrangement is known as 'tip enhanced Raman spectroscopy' (TERS). In combination with SERS conditions, where the specimen is adsorbed onto a nanostructured metallic substrate, the TERS configuration will generate spatially resolved spectra from single molecules.

4.2.5   Technical Implementation and Analytical Methods

The schematic of a Raman confocal spectrometer is shown in Figure 4.2.

The confocal arrangement has the merit of improving the lateral resolution by a factor of two compared with that predicted by the Abbé criterion. Also, it defines the depth resolution, and thus the information volume (see Figure 4.3). Accordingly, it is possible to transfer systematically the interaction volume from the surface into the bulk of the specimen, thus obtaining a depth profile. Indeed, 3-D mapping of a

Confocal Microscopy
3-D Confinement of Measuring Volume
by Confocal Aperture

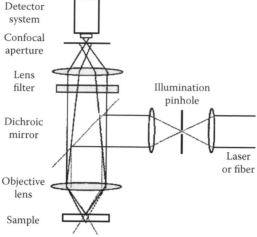

FIGURE 4.2 Schematic diagram of a confocal microscope. Incident light, from a HeNe or Ar ion laser, is filtered by a pinhole, reflected through 90° by a dichroic mirror, and then brought to a focus at a focal plane in the specimen by an objective lens. Elastic and inelastic backscattered light is then focussed onto a confocal aperture before entering the spectrometer and CCD detector. Usually, there is an attenuator in the path of the incident light in order to control the luminosity at the focal point on the specimen. As well, the elastic Rayleigh-scattered peak is filtered out with a notch filter, allowing measurement of Raman shifts down to ca. 100 cm^{-1}. Polarization-dependent data can be obtained by a combination of a polarizer and a quarter wave plate introduced into the optical path. The dispersive element in the spectrometer is generally a grating with 500 to 2000 lines per cm, leading to a spectral resolution from 10 to 1 cm^{-1}, while parallel counting is accomplished with a CCD detector plate. (From Schrof et al., 2001, *Prog. Organic Coatings,* with permission from Elsevier.)

specimen can be carried out with a motorized x-y-z sample stage. In the context of nanostructures, an additional merit of the confocal configuration is suppression of the, usually much stronger, Raman signal originating from a substrate.

4.3 FLUORESCENCE SPECTROSCOPY

The initial step in the fluorescence process is the excitation, by absorption of a photon, from a ground state electronic configuration to an excited state, where the excited state involves an electronic transition and a modified vibrational configuration. The initial relaxation will in general be through one or more of the vibrational states by non-radiative processes (e.g., intermolecular collisions or phonon collisions). Once the relaxation has reached the lowest vibrational state in the excited electronic state, then there is relaxation back to one of the vibrational states of the electronic ground state accompanied by the emission of a fluorescence photon of characteristic energy. Since there may be several available vibrational destination states, there will be a

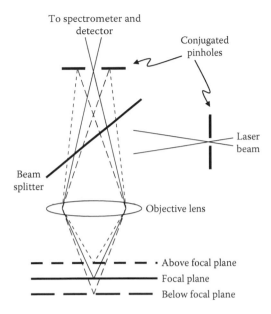

FIGURE 4.3 Illustration of confocal arrangement leading to improved lateral and depth resolution. A 100X objective gives a Raman sampling volume of ca. 1 μm³.

spectrum of fluorescence energies. Accordingly, fluorescence spectroscopy can offer insight into the electronic structure, as well as the vibrational structure, of the specimen. The process couples directly with the vibrational and electronic structure of the sample, (i.e., the intermediate stage of polarisation is not required), and is therefore far more efficient than the Raman process.

4.3.1 Technical Implementation and Operational Modes

In the emission mode, a fluorescence spectrum is generated in response to excitation by light with a wavelength shorter than that required to reach at least the first excited state (i.e., ultraviolet light (UV)). Alternatively, in the excitation mode, a series of emission spectra is generated as a function of wavelength of the incident light. The operation of some instruments is based on filtering of the incident light from a white light source (e.g., a tungsten lamp), or on selecting, with a filter, a particular emission line (e.g., from an Xe lamp or a Hg vapour lamp). More sophisticated instruments may achieve a higher degree of monochromaticity with a grating. A laser excitation source has the advantage of being monochromatic, thus not requiring any further filtering, but the disadvantage of having a fixed wavelength. Similarly, the spectral dispersive element can be a filter, or a grating spectrometer. In the life sciences, the emphasis is generally on tracking a fluorophor with known emission wavelength incorporated into a biosample, and filtering is generally sufficient (see Figure 4.4).

In order to detect fluorescence emission from a single nanostructure with any optical probe, there are two fundamental requirements that a spectroscopic method has to fulfil:

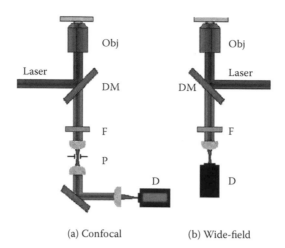

(a) Confocal   (b) Wide-field

FIGURE 4.4 Diagram of the optics of a confocal (a) and a wide-field fluorescence micro-scope (b). The components include: (Obj) microscope objective, (DM) dichroic mirror, (F) filters, (P) pinhole or confocal aperture, and (D) a detector. Dispersive grating elements are not shown. (From Gomez et al., 2006, *Phys. Chem. Chem. Phys.*, with permission from the Royal Society of Chemistry.)

- Only one nanostructure should be present within the interaction volume of the probe at any one time. Accordingly, the structures have to be dispersed on a substrate or in a fluid, so that there is statistical certainty that there is only one structure within a volume of 1 μm^3.
- The chosen technique should be able to detect the relevant scattered radia-tion from the structure with adequate signal-to-noise.

It is essential that components in the optical paths be made from materials that do not contribute to fluorescence. Likewise, any substrate material, or the suspending fluid, must also not contribute.

These and other considerations have been discussed at length in the review litera-ture (see Moerner and Fromm 2003; Michalet and Weiss 2003; Tamarat et al. 2000; Gomez et al. 2006).

A particular variant of fluorescence spectroscopy has been developed for carbon nanotubes (CNT). A schematic is shown in Figure 4.5. Its application to single wall CNTs (SWCNT) will be described in Chapter 9. The particular merit of the method is that it can discriminate on the basis of the chirality of the tube (as defined by (m,n) indices, see Chapter 9).

When the illumination from the tungsten lamp is blocked, a SWCNT sample with mixed chiralities, (dispersed in a fluid), is excited sequentially by diode lasers at three different wavelengths. The corresponding near-IR emission spectra are collected at 180° and detected by a multi-channel array. In order to measure the absorption spec-trum, the lasers are turned off and the tungsten lamp is unblocked. The same detection system records the spectrum of near-IR light transmitted through the SWCNT sample. The result is then compared with the transmission through a blank sample to obtain the

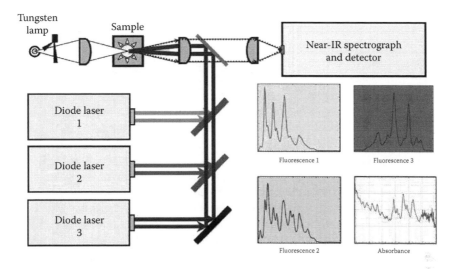

FIGURE 4.5 Schematic diagram of a fluorescence spectrometer that has been optimised for discrimination of SWCNTs with different chiralities. (From Weisman, 2009, *Anal. Bioanal. Chem.*, with kind permission from Springer Science+Business Media.)

absorption spectrum. Each major peak in the resulting excitation emission plot can be associated with a specific (n,m) species (see Chapter 9) with the aid of predetermined tables of SWCNT spectral assignments (Weisman 2008; Weisman 2009).

4.3.2 FLUORESCENCE SPECTROSCOPY—BIOMOLECULAR APPLICATIONS

In dilute concentrations of biomolecular fluorophores, the fluorescence intensity will, in general, be proportional to the degree of dilution. Thus, in principle, semi-quantitative mapping can be carried out. However, several factors conspire to prevent spectra obtained from biomolecular fluorophores being quantifiable. Some of these factors are instrumental, as indicated below.

(a) The intensity of illumination will vary from one source to another, and will in general be a function of wavelength, leading to spectral intensities that are a function of source. This can be corrected by diverting a fixed percentage of the beam, after the monochromator or filter, and then to use this part of the beam as a reference for normalization of the spectra.

(b) The transmission efficiencies of monochromators and filters in the optical path after the scattering will also depend on wavelength, and they need to be taken into account.

(c) The quantum efficiency of the detector, whether single- or multi-channel, will depend on the wavelength of the dispersed fluorescence signal, and the efficiency may indeed change over time.

When only 'finger-print' spectral information is required, such instrumental corrections are rarely carried out. However, when semi-quantitative conclusions based on relative intensities are required, then instrumental factors must be taken into account in order to arrive at a standard spectrum.

Other factors that introduce difficulties into quantification reside in the specimens themselves, viz.,

(i) The nature of the specimen. The most serious factor is the likelihood of photo-decomposition during the period of exposure of the biomolecular fluorophore; the phenomenon is known colloquially as 'bleaching'. Bleaching is, in general, not a problem with the more stable quantum dot structures.

(ii) Reabsorption, which becomes a problem when another biomolecular species in the sample is an absorber in the spectral range over which the fluorophore emits. The effect is exacerbated in thick specimens. Likewise, thick specimens with high concentrations of fluorophores will degrade the fluorescence signal due to reabsorption. These and other factors that affect the quality of fluorescence analysis have been discussed in the literature (Lakowicz 2006; Sharma and Schulman 1999).

4.3.3 SPECTROSCOPIC ANALYSIS OF QUANTUM DOTS

Binary Cd-based nanocrystals are the original and most widely investigated class of quantum dots (Yu et al. 2003). A well-established wet chemical synthesis process allows tight control over the size fraction and phase purity (Yu and Peng 2002; Qu and Peng 2002). The results in Figure 4.6 illustrate that the combination of ultraviolet-visible (UV-Vis) and fluorescence spectroscopies is a convenient method for the determination of the electronic structure as a function of size. Alternatively, the combination can serve as a method for validating size and size distribution.

4.3.4 CARBON-BASED NANOSTRUCTURES AND RAMAN SPECTROSCOPY

The majority of nanostructures of greatest current interest, aside from quantum dots, are carbon-based. The details of their native structural characteristics have been established by high-resolution imaging and analytical S/TEM (see Chapter 2 and Chapter 8), while defect structures and the effects of dopants are works in progress.

The popularity of optical spectroscopies based on Raman and fluorescence scattering predates the age of nanotechnology. In the case of Raman, it became the favoured means for characterizing diamond-like carbon (DLC) products from the various synthesis routes.

There are good reasons for setting the stage for the interpretation of Raman spectra of nanotubes, fullerenes, and graphene, with an overview of the extensive work on bulk graphite-like and diamond-like materials. For instance, a single- or multi-wall CNT can be viewed as the result of the rolling-up of one, or more, graphene layers into a seamless cylinder. Therefore, CNTs are closely related to their parent material, graphite, which exhibits interesting higher-order and defect-induced Raman bands, not usually seen in the typical Raman spectra of solids. Likewise, a fullerene molecule can be viewed as a closed surface, based on deformation of a graphene sheet, constructed from hexagons, pentagons, heptagons, and so forth. A graphene sheet is

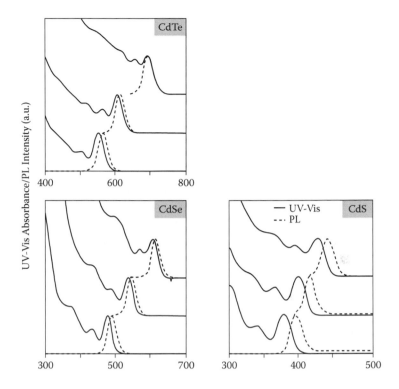

FIGURE 4.6 UV-Vis absorption (solid lines) and fluorescence emission (broken lines) spectra of representative nanocrystal samples of CdTe, CdSe, and CdS. The spectra, left to right, were obtained from size fractions of 18, 25, 29 nm, with a spread of ± 2 nm. (From Yu et al., 2003, *Chem. Mater.*, with permission from the American Chemical Society.)

just a single delaminated layer of graphite. Accordingly, the extensive experimental and theoretical body of knowledge on Raman spectroscopy of carbon compounds will constitute a useful point of departure.

Fluorescence spectroscopy became the favoured tool in biomedical and biomolecular research for tracking compounds and processes labelled by fluorophores, such as green fluorescent protein (GFP) (Tsien 1998).

It is instructive to consider the Raman spectra for the main allotropes of carbon, including diamond, graphite, fullerenes, nanotubes, and graphene.

The common crystalline phases of carbon have comparatively simple spectra. Thus, diamond (with sp^3 hybridisation) has a peak at 1332 cm^{-1} (a single mode with T_{2g} symmetry), whereas graphite (with sp^2 hybridisation) has doubly-degenerate E_{2g} modes at 42 and 1582 cm^{-1}. The former is due to coupling of the anti-parallel vibrations in two adjacent planes with the weak van der Waals inter-plane interaction. Thus, it is a 'soft' mode. It is difficult to access, being very close to the intense Rayleigh line, and is rarely used for characterization. The latter is referred to as the G band and corresponds to vibrations in the graphene planes (see Figure 4.7). The width of the G band depends on crystallographic quality and is a useful diagnostic of extent of disorder.

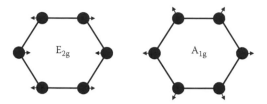

FIGURE 4.7 The two most prominent characteristic modes associated with sp^2 bonding of graphite planes. (Left) The E_{2g} so-called G-mode. (Right) The A_{1g} so-called D breathing mode.

Two additional modes appear whenever flaws are present, the grain size is reduced, or graphene planes are bent (Roy et al. 2003). These modes are called D and D' (signifying disorder). The D' mode arises from the splitting of the G band and is generally found at ca. 1620 cm^{-1}. The ratio of intensities of G to D' bands depends on the proportion of distorted graphene planes, according to the following expression (Solin 1990):

$$\frac{I_G}{I_{D'}} = \left(\frac{n-2}{2}\right)\frac{\sigma_G}{\sigma_{D'}} \tag{4.6}$$

where σ refers to the respective scattering cross-sections and n is the number of graphene layers.

The D peak, found at ca. 1350 cm^{-1}, is a breathing mode of A_{1g} symmetry, shown in Figure 4.7, but is forbidden in perfect graphite, that is, it is absent for a perfect graphene layer, and only becomes active in the presence of disorder. The mode is dispersive, that is, it varies with photon excitation energy (Ferrari and Robertson 2000). It has been shown (Tuinstra and Koenig 1970) that the ratio of the intensity of the D peak to that of the G peak varies inversely with the sp^2 cluster size (a domain size), L_a.

$$\frac{I_D}{I_G} = \frac{C(\lambda)}{L_a} \tag{4.7}$$

where $C(\lambda)$ is a parameter that depends on the wavelength of the incident radiation (e.g., $C \approx 4.4$ nm for $\lambda = 514.5$ nm). However, the intensity and wave number shift appear to be related more generally to imperfections in the sp^2 structure.

A description of the transition from the perfect crystalline graphitic end phase, with 100% sp^2 character, through nanocrystalline (NC) graphite, amorphous carbon (a-C), to tetrahedral amorphous carbon (ta-C), with various ratios of sp^2/sp^3 bonding, and the associated G-peak position and I_D/I_G ratio, has been suggested (Ferrari and Robertson 2000), and is shown in Figure 4.8.

While the D breathing mode is active only in the presence of defects (Tuinstra and Koenig 1970; Ferrari and Robertson 2000; Ferrari and Robertson 2001; Thomsen and Reich 2000), a second-order 2D mode, found at ca. 2700 cm^{-1}, can be a

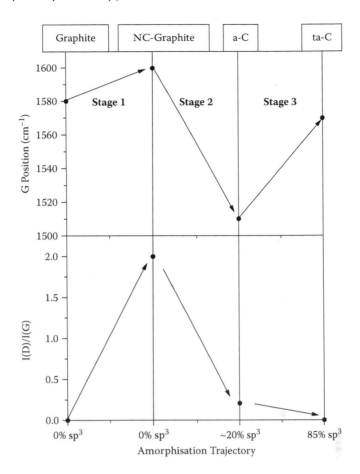

FIGURE 4.8 Trends in Raman signatures for a range from graphite-like (0% sp^3) to diamond-like (85% sp^3) with various degrees of crystallinity and of the ratio of sp^2/sp^3 bonding. (From Ferrari and Robertson, 2000, *Phys. Rev. B*, with permission from the American Physical Society.)

prominent feature, even in the absence of a D peak. In particular, perfect single-layer graphene exhibits a strong and sharp 2D peak; the intensity of the mode is observed to decrease and adopt a more complex structure with the number of graphene layers (Ni et al. 2007; Ferrari et al. 2006).

Extension of phonon modes and selection rules from planar graphene to nonplanar CNTs predicts zero frequencies for the perfectly symmetric radial breathing mode (RBM) of a CNT nanotube, since the RBM arises from the acoustic phonon modes of the graphene layer. However, breaking of the planar symmetry leads to the frequency of the perfectly symmetric RBM for an isolated SWNT being inversely proportional to the nanotube diameter, varying from ca. 100 to 250 cm^{-1} for diameters of 1–2 nm, as was predicted (Jishi et al., 1993; Eklund, Holden, and Jishi 1995). The relationship between diameter, d_{CNT}, and wave number shift is of the form

$$\nu_{RBM} = \frac{A}{d_{CNT}} \tag{4.8}$$

where $A = 248$ cm^{-1} is a constant. Similar relationships have been identified for MWCNTs and for bundles of CNTs (Jorio et al. 2004).

This discussion provides an overview aimed at routine characterization of carbon-based nanostructures. Examples of the characterization of tube structures, graphenes, fullerenes, and QDs can be found in Chapter 9, Chapter 11, Chapter 7, and Chapter 8, respectively. Additional details on the theoretical background and methods can be found in the review literature (e.g., Jorio et al. 2004; Gouadec and Colomban 2007; Wu, Yu, and Shen 2010; Kuzmany et al. 2004; Reich and Thomsen 2004; Ferrari and Robertson 2004).

ACKNOWLEDGEMENTS

It is a pleasure to acknowledge many useful comments and helpful conversations with Dr. C. Johnston.

REFERENCES

Eklund, P. C., Holden, J. M., and Jishi, R. A., 1995, *Carbon*, 33, 959.
Ferrari A. C. and Robertson, J., 2000, *Phys. Rev. B*, 61, 14095.
———, 2001, *Phys. Rev. B*, 64, 075414.
———, 2004, *Phil. Trans. R. Soc. Lond.*, 362, 2477.
Ferrari, A. C., Meyer, J. C., Scardaci, V., Casiraghi, C., Lazzeri, M., Mauri, F., Piscanec, S., Jiang, D., Novoselov, K. S., Roth, S., and Geim, A. K., 2006, *Phys. Rev. Lett.*, 97, 187401.
Gomez, D. E., Califano, M., and Mulvaney, P., 2006, *Phys. Chem. Chem. Phys.*, 8, 4989.
Gouadec, G. and Colomban, P., 2007, *Prog. Cryst. Growth Charact. Materials*, 53, 1.
Jishi, R. A., Venkataraman, L., Dresselhaus, M. S., and Dresselhaus, G., 1993, *Chem. Phys. Lett.*, 209, 77.
Jorio, A., Saito, R., Dresselhaus, G., and Dresselhaus, M. S., 2004, *Phil. Trans. R. Soc. Lond. A*, 362, 2311.
Kneipp, K., Kneipp, H., Itzkan, I., Dasari, R. R., and Feld, M. S., 2002, *J. Phys. Condens. Matter*, 14, R597.
Kuzmany, H., Pfeiffer, R., Hulman, M., and Kramberger, C., 2004, *Phil. Trans. R. Soc. Lond. A*, 362, 2375.
Lakowicz, J. R., 2006, *Principles of Fluorescence Spectroscopy*, 3rd ed., NY: Springer.
Landau, L. D. and Lifshitz, E. M., 1960, *Mechanics*, Oxford: Pergamon.
Lewis, I. R. and Edwards, H. G. M., (eds.), 2001, *Handbook of Raman Spectroscopy: From the Research Laboratory to the Process Line*, New York: Marcel Dekker.
Lin-Vien, D., Colthup, N. B., Fateley, W. G., and Grasselli, J. G., 1991, *Handbook of Infrared and Raman Characteristic Frequencies of Organic Molecules*, San Diego: Academic Press.
Michalet, X. and Weiss, S., 2002, *C. R. Phys.*, 3, 619.
Moerner, W. E. and Fromm, D. P., 2003, *Rev. Sci. Instrum.*, 74, 3597.
Moskovits, M., Tay, L. L., Yang, J., and Haslett, T., 2002, SERS and the Single Molecule, In Ozin, G. A., Arsenault, A. C., and Cadermarlioi, L., (eds.), *Optical Properties of Nanostructured Random Media*, Vol. 82, p. 215–226. Berlin: Springer.

Nakamoto, K., 2009a, Infrared and Raman Spectra of Inorganic and Coordination Compounds: Theory and Applications, In: *Inorganic Chemistry*, 6th ed., Part A, Hoboken, NJ: Wiley.

————, 2009b, *Infrared and Raman Spectra of Inorganic and Coordination Compounds: Part B, Applications in Coordination, Organometallic, and Bioinorganic Chemistry*, 6th ed., Hoboken, NJ: Wiley.

Ni, Z. H., Wang, H. M., Kasim, J., Fan, H. M., Yu, T., Wu, Y. H., Feng, Y. P., and Shen, Z. X., 2007, *Nano Lett.*, 7, 2758.

Nie, S. M. and Emory, S. R., 1997, *Science*, 275, 1102.

Otto, A., 2002, *J. Raman Spectrosc.*, 33, 59.

Qu, L. and Peng, X., 2002, *J. Am. Chem. Soc.*, 124, 2049.

Reich, S. and Thomsen, C., 2004, *Phil. Trans. R. Soc. Lond. A*, 362, 2271.

Roy, D., Chhowalla, M., Wang, H., Sano, N., Alexandrou, I., Clyne, T. W., and Amaratunga, G. A. J., 2003, *Chem. Phys. Lett.*, 373, 52.

Schrof, W., Beck, E., Etzrodt, G., Hintze-Brüning, H., Meisenburg, U., Schwalm, R., and Warming, J., 2001, *Prog. Organic Coatings*, 43, 1.

Schwartz, D. T., *Raman Spectroscopy: Introductory Tutorial*, Seattle, WA: University of Washington.

Sharma, A. and Schulman, S. G., 1999, *Introduction to Fluorescence Spectroscopy*, New York, NY: Wiley Interscience.

Solin, S. A., 1990, Graphite Intercalation Compounds I, Ch. 5, In *Structure and Dynamics*, Vol. 157, Zabel, H. and Solin, S. A., (eds.), Berlin: Springer-Verlag.

Tamarat, P., Maali, A., Lounis B., and Orrit, M., 2000, *J. Phys. Chem. A*, 104, 1.

Thomsen, C. and Reich, S., 2000, *Phys. Rev. Lett.*, 85, 5214.

Tsien, R., 1998, *Ann. Rev. Biochem.*, 67, 509.

Tuinstra, F. and Koening, J. L., 1970, *J. Chem. Phys.*, 53, 1126.

Turrell, G., 1972, *Infrared and Raman Spectra of Crystals*, London, UK: Academic Press.

Turrell, G. and Corset, J. (eds.), 1996, *Raman Microscopy: Developments and Applications*, San Diego, CA: Elsevier.

Weisman, R. B., 2008, *Contemp. Concepts Cond. Matter Sci.*, 3, 109.

Weisman, R. B., 2009, Fluorimetric Characterization of Single-Walled Carbon Nanotubes. *Anal. Bioanal. Chem.*, 396, 1015.

Wu, Y. H., Yu, T., and Shen, Z., X., 2010, *J. Appl. Phys.*, 108, 071301.

Yu, W. W. and Peng, X., 2002, *Angew. Chem., Int. Ed.*, 41, 2368.

Yu, W. W., Qu, L., Guo, W., and Peng, X., 2003, *Chem. Mater.*, 15, 2854.

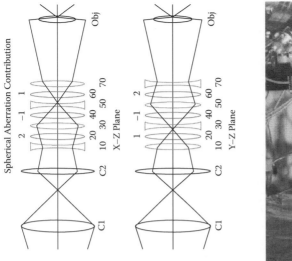

FIGURE 2.8 The Nion-designed aberration corrector for an STEM instrument, consisting of a stack of quadrupole and octupole lenses between the objective and condenser lenses. (From IBM™ Web site, also see Dellby et al., 2001.)

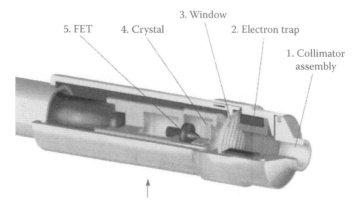

FIGURE 3.7 The energy-dispersive EDS detector. (From http://www.charfac.umn.edu/instruments/eds_on_sem_primer.pdf.)

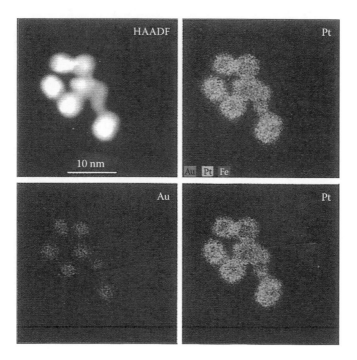

FIGURE 3.10 Compositional mapping by EDS of particles in the 5 nm size range. (From Schlossmacher et al., 2010, *Microscopy and Analysis*, November, p. 55. With permission.)

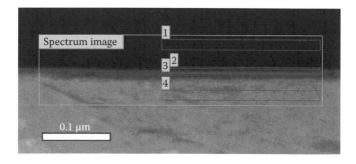

FIGURE 3.15 The four regions of interest selected for EELS and EDS analyses are shown. (From Kundmann, 2007, http://www.gatan.com/resources/answers/1.php.)

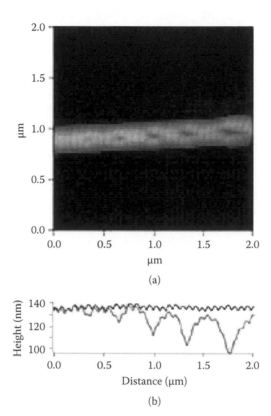

(a)

(b)

FIGURE 5.21 (a) Plastic deformation of a collagen fibril resulting from indentation by AFM for loads up to 10 μN. (b) Contour line along the centre of the fibril in (a) shows indentation pits of increasing depth and width. The black line corresponds to the un-deformed fibril, with the corrugations showing the periodicity of the so-called D-band spacing. The image was obtained in the tapping mode. (Adapted from Grant et al., 2009, *Biophys. J.*, with permission from Elsevier.)

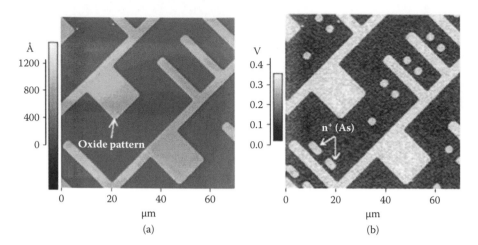

(a)   (b)

FIGURE 5.27 Contact mode topographic (left), and SCM (right), images of a patterned semiconductor surface. Regions with bright contrast in the topographic image represent a thermally grown SiO_2 pattern of height 70 nm. The additional circular and rounded rectangular features evident in the SCM image correspond to regions heavily doped by ion implantation (50 keV As^+ ions and 10^{14} ions/cm² dose density). (From Yoo et al., Park Systems Web site, SCM Mode Note, www.parkafm.co.kr. With permission.)

FIGURE 5.32 Shear force and optical transmission images, left and right, respectively, of a standard SNOM specimen. The specimen was produced by metal evaporation onto a flat substrate with close-packed polystyrene spheres as a shadow-mask (the spheres were subsequently removed). (With kind permission from D. Higgins, 2007, Kansas State University.)

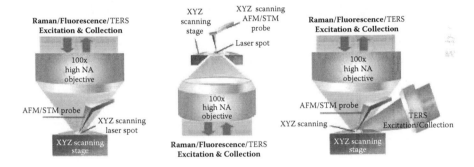

FIGURE 5.34 Possible TERS configurations. (Left) A high numerical aperture objective brings the incident radiation to a focus at the tip apex. The diffraction-limited resolution of the objective is ca. 400 nm in air. The configuration is appropriate for AFM imaging and spectral analysis of opaque samples. (Middle) This configuration is based on an inverted microscope base, and is appropriate for spectral analysis in transmission with a nanoscale specimen deposited on a microscope slide. (Right) Here the illumination is incident from the side, with scattered radiation collected in the backscatter mode. However, the arrangement can also be reversed, with vertically incident illumination, and scattered radiation collected from the side (at the expense of a smaller solid angle of collection). (Adapted from Dorozhkin et al., 2010, *Microsc. Today,* November, p. 28.)

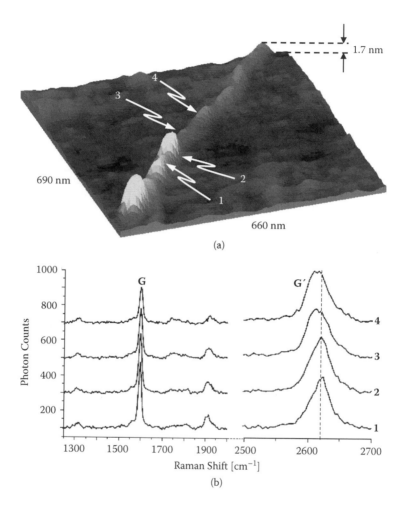

FIGURE 5.35 (Top) Topographic image, shown in 3-D representation, of a SWCNT, grown by arc discharge, on a glass substrate. The apparently excessive width of the CNT was due to tip shape convolution. (Bottom) Near-field Raman spectra recorded at positions 1 to 4 in the image. The spacing between positions 2 and 3 was ca. 35 nm. The spectra are offset in the interest of clarity. (Adapted from Hartschuh et al., 2003, *Phys. Rev. Letts.*)

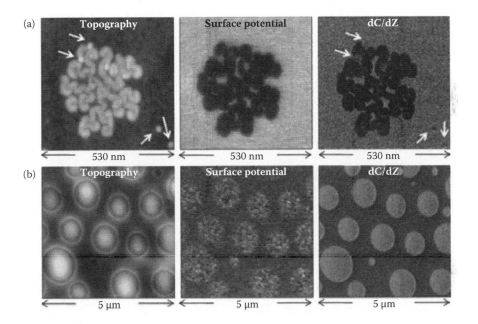

FIGURE 5.39 Row A shows images of a fluoroalkane adsorbate ($F_{14}H_{20}$) on a Si substrate. The ranges of contrast are 0–7 nm for topography, 0–1.1 V for surface potential, and 0–0.15 V for dC/dz. Row B shows images for a 70 nm thick film of polystyrene/poly(vinyl acetate) 1:1 blend on an ITO (indium-tin oxide) substrate. The ranges of contrast are 0–28 nm for topography, 0–0.6 V for surface potential, and 0–0.11 V for dC/dz. (From Magonov and Alexander, 2010, *G.I.T. Imaging Microscopy*. With permission from S. Magonov.)

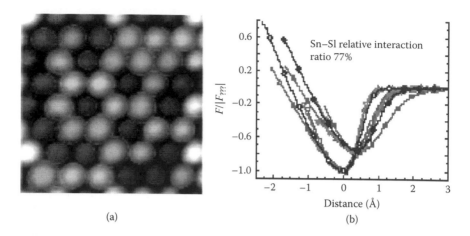

(a)   (b)

FIGURE 5.44 (Left) False colour image of a surface alloy composed of Si (red-brown), Pb (green), and Sn (blue) atoms deposited on an Si(111) surface. The field of view is 4.3×4.3 nm². (Right) Normalized F-d curves showing the distinct difference in interaction between the tip and Si and Sn surface atoms (the corresponding results for Pb-Si and Si-Si interactions are not shown). Curves of different colour refer to data obtained with different probes, and they demonstrate that the assignments are independent of the tip material. The F-d data have been normalised to those recorded for the Si species ($F/F_{Si\text{-}set}$ = 1 at distance = 0). (Adapted from Sugimoto et al., 2007, *Nature*, with permission from Macmillan Publishers Ltd.)

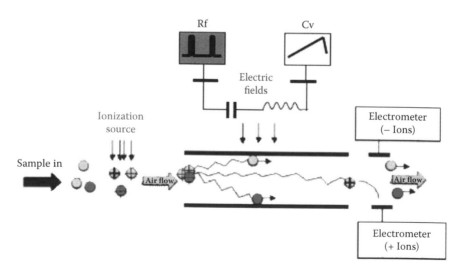

FIGURE 6.14 Schematic of a differential mobility spectrometer (DMS). (Adapted from Davis, 2009, *Intl. Gases Instrum.*)

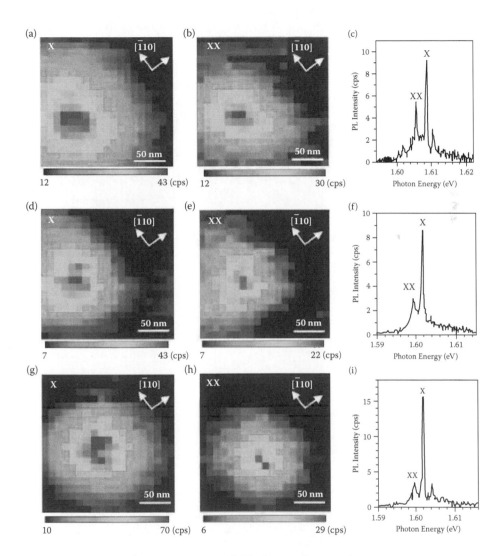

FIGURE 8.18 The data refer to three individual quantum dots, where the exciton decay proceeds through two channels, giving rise to emission lines labelled X and XX. The intensity of emission line X is mapped for the three dots in a, d, and g, while the XX line is mapped in b, e, and h. The integrated intensities for the three dots are shown in spectral form in c, f, i. (From Matsuda et al., 2003, *Phys. Rev. Lett.*, with permission from the American Physical Society.)

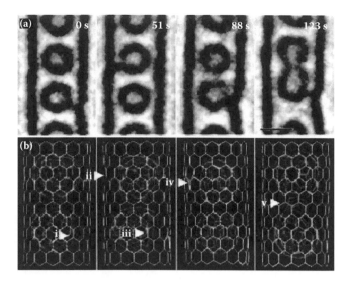

FIGURE 9.14 HRTEM imaging of C_{92} molecules inside a SWCNT; effects of electron irradiation dose. (From Urita et al., 2004, *Nano Lett.*, with permission from the American Chemical Society.)

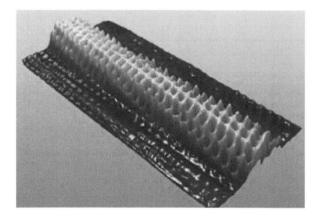

FIGURE 9.23 STM image of a chiral CNT; the diameter of the tube was ca. 1 nm. (From the Image Cees Dekker Group at TU Delft, The Netherlands, reproduced with permission from the author.)

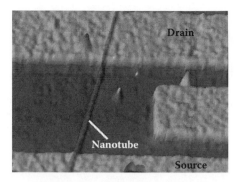

FIGURE 9.24 An AFM image showing an FET device constructed by bridging source and drain electrodes by a single CNT. The SiO$_2$ substrate constitutes the gate dielectric. (From Martel et al., 1998, *Appl. Phys. Lett.*, with permission from the American Institute of Physics.)

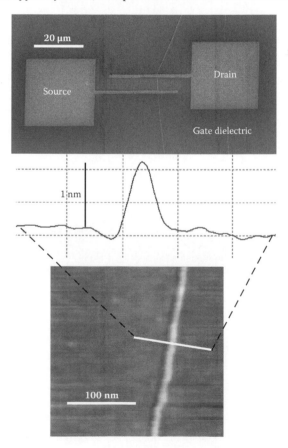

FIGURE 9.25 Illustration of SEM and AFM characterization of a CNT-based device. The SEM image (top) shows the lay-out of the device, while the middle AFM line trace, taken from the AFM tapping mode image (bottom), demonstrates that the CNT was single wall. (From Inzani, 2011, unpublished data, University of Oxford, with permission.)

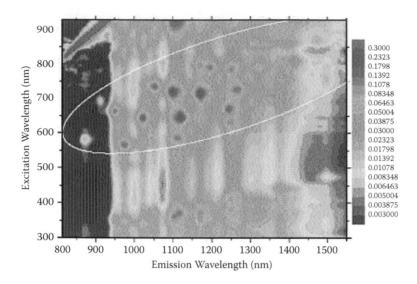

FIGURE 9.34 False-colour contour plot of the data in Figure 9.33 shows the precise wavelengths for each peak. (From Weisman, 2008, *Contemporary Concepts of Condensed Matter Science*, with permission from Elsevier.)

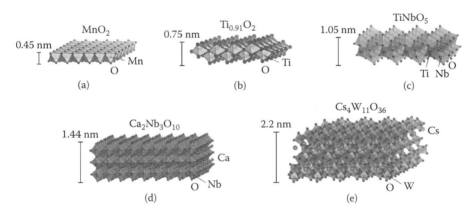

FIGURE 11.3 Examples of oxide nanosheets. (a) MnO_2, (b) $Ti_{1-\delta}O_2$ (δ denotes Ti deficiency), (c) $TiNbO_5$, (d) $Ca_2Nb_3O_{10}$, (e) $Cs_4W_{11}O_{36}$. (From Osada and Sasaki, 2009, *J. Mater. Chem.*, reproduced with permission from the Royal Society of Chemistry.)

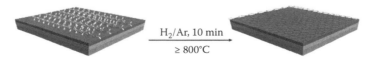

FIGURE 11.4 Schematic of the process route for catalyst-promoted, substrate-mediated synthesis of graphene. (Adapted from Sun et al., 2010.)

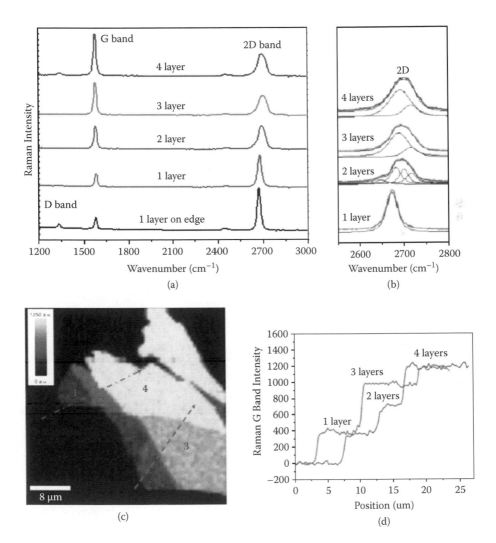

FIGURE 11.12 (a). Raman spectra as a function of number of layers. (b) Details of the structure of the 2D band. The added complexity of the 2D structure is due to intralayer interactions, which perturb the in-layer modes. (c) Raman image generated by the intensity of the G band. (d) Cross sections of Raman image along the broken lines in (c). (From Ni et al., 2007, *Nano Lett.*, with permission from the American Chemical Society.)

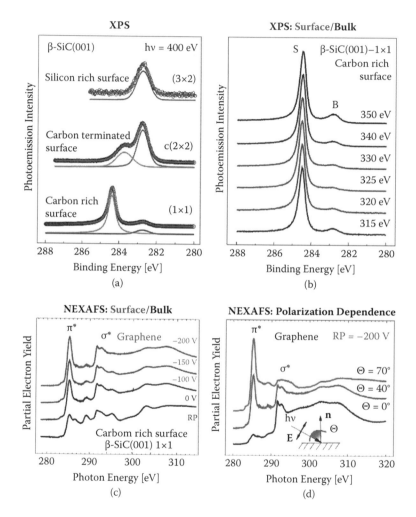

FIGURE 11.14 Illustration of the basic characterization by synchrotron photoelectron spectroscopy of the electronic structure of graphene synthesised on β-SiC substrate. (From Aristov et al., 2010, *Nano Lett.*, with permission from the American Chemical Society.)

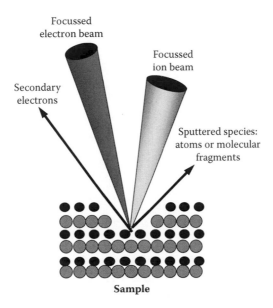

FIGURE 12.5 Schematic depiction of the dual beam arrangement. Incident ions sputter away atoms or molecular fragments from the point of focus. Secondary electrons are ejected from the respective focal points of the electron and ion beams.

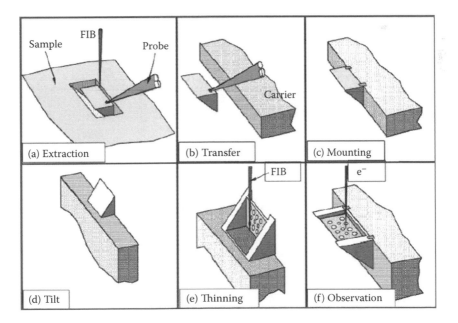

FIGURE 12.8 Schematic illustrations of the stages in the extraction, transfer, mounting, and final preparation by the wedge method of a sample for TEM analysis, by the INLO technique. (From Mayer et al., 2007, *MRS Bull,* from the Cambridge University Press, with permission.)

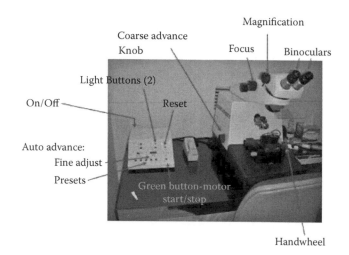

FIGURE 12.9 A typical microtome. The knife and sample holder are viewed through a binocular microscope. (From http://sbs.wsu.edu/fmic/protocol/microtome.ppt.)

5 Scanning Probe Techniques and Methods

5.1 INTRODUCTION

The first members of the scanning probe microscope (SPM) family, the scanning tunnelling microscope (STM) and the atomic force microscope (AFM), were conceived and developed in the early to mid-1980s (Binnig and Rohrer 1982; Binnig, Quate, and Gerber 1986). Since then the technologies and ideas that underpinned the original developments have provided the inspiration for numerous additional related techniques, each capable of functioning in several different operational modes. Until recently, SPM was considered to be a novel group of techniques, in comparison with the venerable electron-optical microscopic family. However, even though the entire spectrum of variations on the SPM theme is still being explored, and the gradual evolution of instrumental technology remains a work in progress, SPM must now be considered as a mature technique, with well-established protocols, strengths, and weaknesses. Thus, the present chapter will refrain from describing the technical intricacies of an SPM instrument, but instead will refer interested readers to the relevant literature (e.g., Magonov and Whangbo 2007; DiNardo 2007; Meyer, Hug, and Bennewitz 2004; Wiesendanger, Meyer, and Morita 2002; Wiesendanger 1998; Myhra 2010). Instead, it will address the many operational modes and related methods that are of particular relevance to the characterization of nanostructures.

5.1.1 ESSENTIAL ELEMENTS OF SPM

An SPM system, shown schematically in Figure 5.1, consists of a surface having structural and physicochemical attributes, a tip, also with structural and physicochemical attributes, that can be located and controlled in the spatial and temporal domains, and interactions with characteristic strengths and ranges that depend on the respective attributes of surface and tip. The latter allows a two-way transfer of 'information' between tip and surface. On the basis of this description, the observation can be made that, given sufficient knowledge about any two of the three elements, in principle, knowledge can be gained about the third element. Therein is the richness of information inherent in SPM techniques.

An abbreviated tree of the SPM family is shown in Figure 5.2.

The salient characteristics of some members of the SPM family are sketched in Figure 5.3. In the case of SThM, the tip is essentially a local thermal radiation sensor (i.e., a thermocouple or a thermistor). The underlying mechanism of STMIP is that of the emission of low energy photons in response to tunnel electrons being injected

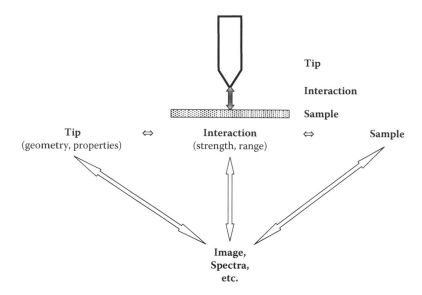

FIGURE 5.1   Schematic representation of the essential elements of an SPM system. (From Myhra, S., 2010, in *Handbook of Surface and Interface Analysis*, Rivière, J. C. and Myhra, S., eds., with permission from Taylor & Francis.)

into unoccupied surface states. SICM is based on the sensing of ion current being drawn through a nanocapillary tube in close proximity to a surface.

Nearly all the traditional techniques for the characterization of materials have emerged from the exploitation of new technologies in combination with significant scientific discoveries, for example, X-ray diffraction from the discovery of X-rays, surface spectroscopies from the then mature field of atomic spectroscopy in combination with ultra-high vacuum (UHV) technology, Raman spectroscopy from the discovery of the Raman effect in combination with laser light sources, and so forth. SPM, on the other hand, arrived unannounced by any noteworthy precursor discovery or technology. Nevertheless, the family of local probe techniques instantly caught both popular and professional imagination, and became mainstream in less than a decade. Some of the reasons are:

5.1.1.1   Cost-Effectiveness

The capital cost of a multi-technique, multi-mode SPM instrument is a small fraction of that of an UHV-based surface technique or a state-of-the-art analytical SEM or TEM. The exception is an UHV-STM facility. Likewise, the day-to-day cost of operation of SPM techniques compares favourably, the principal cost being that of the probes themselves.

5.1.1.2   Platform Flexibility

An SPM instrument is essentially a tool-kit that can be configured readily to support the great majority of local probe techniques and their respective operational modes. Other techniques can be added to an existing platform with modest capital outlays.

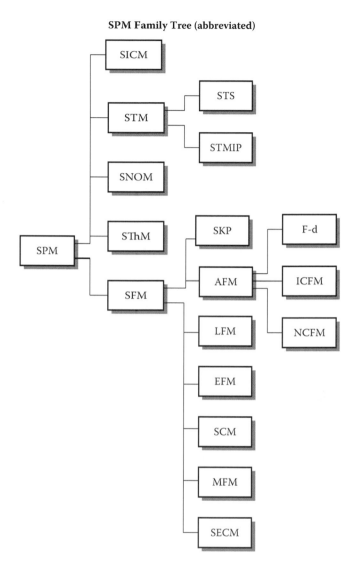

SPM Family Tree (abbreviated)

FIGURE 5.2 The SPM family tree. Acronym definitions: STS = scanning tunnelling spectroscopy; STMIP = scanning tunnelling microscopy inverse photoemission; SNOM = NSOM = scanning near-field optical microscopy; SThM = scanning thermal microscopy; SFM = scanning force microscopy; F-d = force versus distance (analysis); ICFM = intermittent contact (= 'tapping') force microscopy; NCFM = non-contact force microscopy; LFM = lateral force microscopy (also known as scanning chemical force microscopy); EFM = electrostatic force microscopy; SCM = scanning capacitance microscopy; MFM = magnetic force microscopy; SECM = spreading electrical current microscopy; SICM = scanning ion current microscopy; SKP = scanning Kelvin probe. **Reader beware**—other writers may use different acronyms! (Adapted from Myhra, S., 2010, in *Handbook of Surface and Interface Analysis*, Rivière, J. C. and Myhra, S., eds., with permission from Taylor & Francis.)

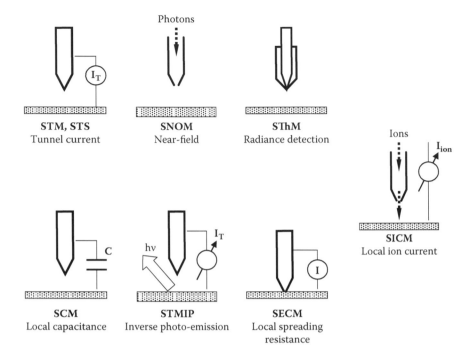

FIGURE 5.3 Schematic description of some SPM techniques. (Adapted from Myhra, S., 2010, in *Handbook of Surface and Interface Analysis*, Rivière, J. C. and Myhra, S., eds., with permission from Taylor & Francis.)

5.1.1.3 Ambient Tolerance

One of the great strengths of SPM techniques resides in the relative insensitivity of performance to almost any ambient environment, ranging from bio-compatible fluids, via either laboratory ambient or a controlled gaseous environment, through to high and ultra-high vacuum.

5.1.1.4 User-Friendliness

Analysis in ambient or controlled atmospheres is consistent with minimal specimen preparation and rapid turn-around, thus being comparable to SEM, but as distinct from the preparation of TEM foils. Obtaining a usable AFM image in the contact mode requires only marginally greater training than that for a standard SEM; mounting and aligning the probe is probably the most taxing task.

5.1.1.5 Ease of Interpretation

SPM images have the great advantage of providing information in 'real' space, as opposed to techniques that are based on diffraction and/or phase contrast. However, it may take both experience and insight to recognize image artefacts. In the case of F-d analysis, the nature of the information is readily apparent, although its relationship to mechanical or electronic properties may require considerable theoretical insight.

5.1.1.6 Unique Capabilities

The SPM family has had a great impact on recent progress in nanoscience and technology due to its broad range of capabilities, some of which are unique. True 'real' space single-atom and single-molecule resolution for STM and SFM, respectively, was the initial focus of attention. Subsequently, SPM has provided unique opportunities for studies in nanotribology, nanobiology, and nanomechanics, and has provided the means for simultaneous manipulation and visualisation of systems on the meso- and nanoscale.

5.2 TECHNICAL IMPLEMENTATION

A functional block diagram of a generic SPM system is shown in Figure 5.4. There are several essential elements that are common to all instrumental SPM platforms, and which merit brief descriptions.

5.2.1 SPATIAL POSITIONING AND CONTROL

Spatial positioning and control of the probe with respect to a location on the surface, in combination with stability and temporal response, is a key factor in the

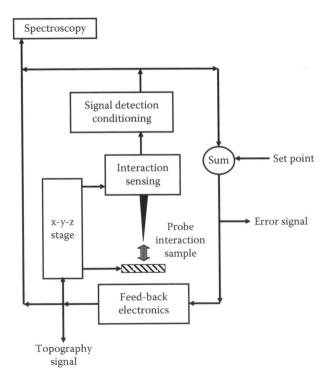

FIGURE 5.4 Functional block diagram of a generic SPM system.

implementation of SPM techniques. Three spatial regimes spanning micro- to nano-scale positioning need to be considered, viz.,

- **Coarse lateral movement** in order to be able to locate promising regions of the surface within the dynamic range of the available field of view. The present generation of instruments achieves that objective either with step-per-motors driving differential screw arrangements, or with an 'inch-worm' arrangement (An inch-worm drive is a piezo-electric device consisting of an expandable body and two clamps).
- **Coarse and fine movements in the z-direction** of the tip, with respect to the surface, in order to bring the probe into the dynamic range of the gap control loop (the tip approach sequence). A stepper-motor method is the most widely used. All commercial instruments have placed the final approach under software control.
- **Lateral rastering** of the tip relative to the surface either by displacement of the sample with respect to a stationary tip (the most common, but not exclusively so, arrangement for SFM instruments), or vice versa (usually the case for UHV-STM instruments).

Piezo-electric materials have the property that a dimensional change is proportional to the local **E**-field. In the case of a beam (or a tube), the change in length is given by

$$\Delta x = \frac{d_{31} V \ell}{h} \tag{5.1}$$

where d_{31} is the piezo-electric coefficient, V is the applied potential between two electrodes, ℓ is the length in the direction of expansion, and h is the spacing between the electrodes (usually the smallest dimension of a beam, or the wall thickness of a tube). Other figures of merit relate to mechanical eigenmodes of the devices, coefficient of thermal expansion, creep, hysteresis, linearity, orthogonality, and aging. The two devices most widely used for positional control, the tube scanner and the tripod, are shown in Figure 5.5. The tube scanner has four external segmented electrodes and a common electrode inside. The tube can undergo either lateral bending, or extension/contraction in the z-direction. Non-ideal characteristics can mostly be overcome by clever engineering, by software correction routines, and by periodic calibration. Routine calibration is usually carried out using semiconductor grids of known dimensions on the μm-scale in the x-y plane and along the z-direction. Suitable standards can now be obtained from most of the manufacturers of SPM instrumentation. On the nanoscale, a freshly cleaved face of highly oriented pyrolytic graphite (HOPG) is a widely used and convenient standard.

Some instruments are based on single scanners with fields of view in excess of 100×100 μm^2, that can be subdivided by digital-to-analogue (D/A) converters and low-voltage modes to achieve bit-size resolutions of 0.01 and 0.1 nm, in the z- and x/y-directions, respectively. Other instruments have exchangeable stages that cover the high and low spatial resolution ranges.

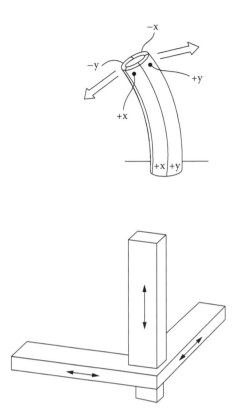

FIGURE 5.5 Schematic illustration of the two most widely used positional devices—the tube scanner (top) and the tripod scanner (bottom). The electrodes on the tube scanner usually consist of, firstly, a continuous conductive coating on the inside of the tube, and, secondly, four longitudinal segments of opposing pairs on the outside. For the tripod scanner, each arm will have two opposing electrodes on the largest parallel faces in the direction of expansion/contraction. (From Myhra, S., 2010, in *Handbook of Surface and Interface Analysis*, Rivière, J. C. and Myhra, S., eds., with permission from Taylor & Francis.)

5.2.2 THE FEEDBACK CONTROL ELECTRONICS

The feedback control electronics close the loop consisting of sample surface, interaction and tip. The essential objective is to ensure the most accurate tracking by the tip of a contour (e.g., of constant tunnel current, attractive force, etc.,) in space, where the contour is defined by a set-point. In order to appreciate fully the power of the technique, the phase shifts and gains in the mechanical components in the loop and the functional dependence of the interaction on gap dimension, should be borne in mind. The loop must thus contain electronics not only for filtering, proportional gain, integral and differential time constants and phase shifting, but also for generating and mixing in other signals for spectroscopy, F-d analysis, tapping mode imaging, and so forth. A general block diagram of the electronics for an STM configuration is shown in Figure 5.6. The parameters and functions are under software

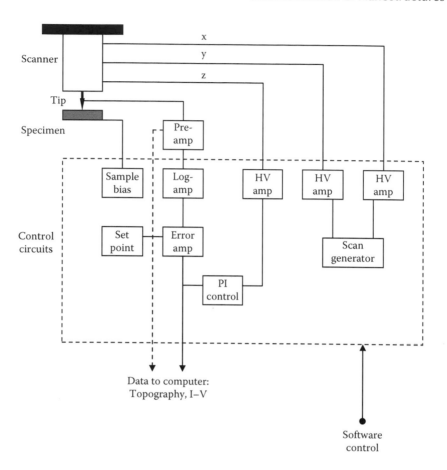

FIGURE 5.6 Block diagram lay-out for a typical STM configuration. The purpose of the log-amp is to linearise the tip current, which has an exponential dependence on the tunnel gap distance. The error amplifier senses deviation from the set-point, and then provides the input for the PI controller (proportional and integral treatment of the error signal). A scan generator drives the raster via high voltage amplifiers. The tunnel gap is maintained by the high voltage input to the z-stage scanner in response to the output from the PI controller. (From Myhra, S., 2010, in *Handbook of Surface and Interface Analysis*, Rivière, J. C. and Myhra, S., eds., with permission from Taylor & Francis.)

control, but are accessible to the experimenter so that optimum conditions can be set up. The quality of the data is related critically to the choice of these variables. Current generation instruments tend to use digital techniques supported by software and A/D and D/A interfacing for instrument control and data acquisition.

The block diagram in Figure 5.7 shows the typical lay-out of SFM control electronics in more detail, and Figure 5.8 shows the arrangement whereby the position-sensitive photo-diode detector (PSPD) can sense the response of a force-sensing lever to out-of-plane and in-plane force components.

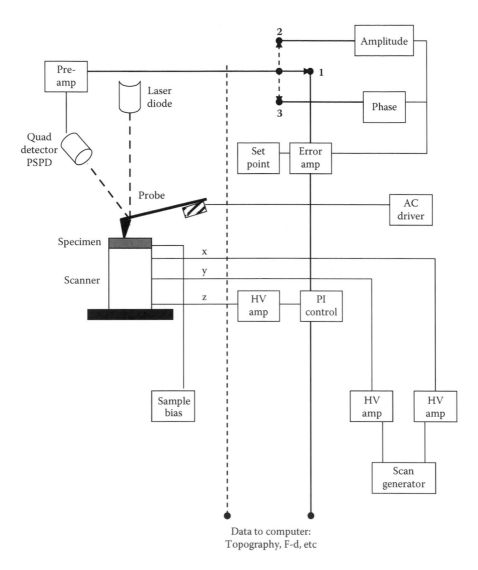

FIGURE 5.7 Block diagram lay-out for a typical SFM configuration. Output from a laser diode is incident on the lever at the location of the tip (tip-plus-lever constitutes the probe); the reflected light is sensed by a position-sensitive photo-diode (PSPD). The resultant optical lever arrangement has an angular deflection sensitivity in the range 10^{-6}–10^{-7} rad. The lever can be stimulated to oscillate at, or near, its fundamental eigenmode by an AC signal driving a piezoelectric actuator at the point of attachment of the lever, thus allowing operation in the non-contact or intermittent contact modes. The latter is also known as the tapping mode. (From Myhra, S., 2010, in *Handbook of Surface and Interface Analysis*, Rivière, J. C. and Myhra, S., eds., with permission from Taylor & Francis.)

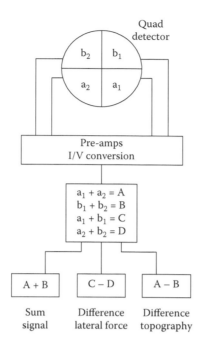

FIGURE 5.8 Signals from the quadrants of the PSPD are manipulated by an arithmetic circuit. The sum signal (A+B) is used to optimise the intensity of the reflected light from the lever during the alignment procedure. In the topographic imaging mode the (A–B) signal is compared to a set-point offset, thus generating an error signal; the latter constitutes the input to the control loop. In the F-d mode, the (A–B) signal is proportional to the normal deflection of the lever, and is thus a measure of the strength of the out-of-plane force component acting on, and imposed by, the tip. The (C–D) signal represents a measure of the torsional distortion of the lever in response to in-plane force components perpendicular to the long axis of the lever acting on the tip. (From Myhra, S., 2010, in *Handbook of Surface and Interface Analysis*, Rivière, J. C. and Myhra, S., eds., with permission from Taylor & Francis.)

5.2.3 THE PROBE

All the science accessible to SPM analysis takes place at the interface between the probe tip and the surface. Accordingly, the choice of probe and its mechanical and physico-chemical properties are crucially important. Specific requirements will be described in subsequent sections, but it is useful to summarise the main requirements.

5.2.3.1 STM

The probe, with a length of a few mm, is prepared from Pt/Ir or W wire, typically of 0.3 mm diameter. A sharp apex is normally produced by electro-chemical etching (see Hansma and Tersoff 1987 for an early description; similar recipes can be found on the Internet). If the surface to be analysed is atomically flat, then the aspect ratio is not relevant, and reliance can be placed on the fact not only that the tunnelling probability will be significant only between the closest pair of atoms, but that there will be one atom at the apex of the probe that will protrude beyond all the others.

That atom will *de facto* be the apex. If the surface is corrugated with steep slopes, then the overall aspect ratio will be important, in order to maximise access by the tip to all parts of the surface. Contamination and oxide, in the case of a W tip in UHV, must be removed, usually *in situ* by field emission, in order to eliminate any surface barrier, and to ensure stability of the apex.

5.2.3.2 SFM

Several factors, additional to those that are relevant for STM probes, affect the choice and performance of SFM probes, and are discussed below. The critical parameters, and the issues that dictate their choice include:

The **normal force constant**, k_N, refers to deflection of the lever due to force components normal to the surface. It must be matched to the operational mode and to the effective force constant of the surface being probed. 'Soft' surfaces (e.g., organics, polymers, biomembranes, etc.,) will be deformed and/or damaged unless a 'soft' lever is being used ($k_N < 0.1$ N/m). On the other hand, contact mode imaging of 'hard' surfaces tends to produce better outcomes with a 'stiff' lever (> 1 N/m). AC mode imaging requires a much stiffer lever, typically with $k_N = 3$–10 N/m, in order to have a principal resonance frequency above 100 khz. The lateral torsional force constant, k_T, is relevant for lateral force imaging and for nanotribological analysis. Beam-shaped probes have simpler bending modes, and are therefore preferred. Sensitivity to in-plane force components is enhanced by high ratios of tip length to length of lever, and by low normal force constant (see below). On the other hand, a V-shaped probe with a short tip is preferable if it is desirable to suppress torsional response.

The **resonance frequency** needs to be no lower than that of the instrumental eigenmode in order to ensure decoupling from mechanical noise (preferably above 10 khz), and in order to allow high scan rates. For AC mode operation, a high resonance frequency ensures a better Q-value (Q = quality) for the resonance envelope and thus, better response of the control loop. (The envelope is essentially a plot of vibrational amplitude of the lever, at constant excitation power, as a function of excitation frequency. The envelope will have a maximum at the eigenfrequency of the lever. Its width depends on dissipation and is measured by the Q-value, defined by the peak frequency divided by the full width-at-half-maximum amplitude.)

The **length of the lever** affects the sensitivity of the optical beam method of detection, which is proportional to $\Delta z/L$, where Δz is the deflection of the lever and L is its length. A lever length of 100 μm will result in z-resolution of better than 0.1 nm for an optical path length, from the lever to the detector, of 1 cm. Lengths of levers range from 100 to 300 μm, with the other dimensions being ca. 30 μm width and ca. 1 μm thickness.

The **optical reflectivity** of the top face of the lever is important in order to improve the signal-to-noise ratio of reflected light intensity, as well as to suppress diffuse scattered stray light, and to ensure that the thermal stability of the assembly is not affected by absorbed energy. Thus, levers are often coated with a thin layer of Al or Au.

The **aspect ratio of the tip** (A_r, height to half-width at base) determines the extent to which high aspect ratio features on a surface can be accessed by the tip. Also, a high aspect ratio will minimize the effects of tip-to-surface 'convolution' in the image. Routine probes may have tips of pyramidal shape with A_r of order unity. Sharpened tips may have A_r in the region of 10 near the apex, while special purpose tips with attached carbon whiskers, or carbon nanotubes, will have even greater A_r values (although they are rather more expensive).

The **radius of curvature of the tip**, R_T, at the apex is an important parameter that will affect the image quality. Any feature in a surface with a radius of curvature less than that of the tip will produce an image of the tip (known as 'reverse imaging' (Hellemans, Waeyaert, and Hennau 1991)). Non-standard tip shapes are also likely to produce image artefacts that can lead to misleading or wrong conclusions. Routine tips may have an R_T of ca. 50 nm, while tips for higher resolution can have R_T values smaller by a factor of 10.

The **surface chemistry of the tip** is of critical importance for F-d analysis (e.g., measurement of adhesion, investigation of biomolecular bonding, protein unfolding, analysis of double layer interactions, etc.). As well, the surface chemistry will affect friction measurements in the LFM mode. An as-received clean tip, microfabricated from doped Si or nominally stoichiometric Si_3N_4, will be covered by native Si-oxide, and will thus present a hydrophilic surface. However, atmospheric exposure renders the tip hydrophobic due to adsorbed hydrocarbon contamination. There is increasing interest in studies requiring specific and known tip functionalities. These can be engineered through silane coupling to the oxide layer (e.g., Plueddemann 1991), or through thiol coupling to an Au coating (e.g., Ulman 1998). Several commercial suppliers offer AFM probes with tailored surface chemistries.

Special purpose probes are required for some applications and particular operational modes. For instance, an MFM probe is commonly produced by magnetising a thin coating of Co on a standard tip. SCM analysis requires that the tip be a 'metallic' conductor. SECM needs to make an ohmic contact with the surface being probed. SThM scanning is carried out with a tip that is essentially a thermometer (either a thermocouple junction or a thermistor device).

5.2.3.3   SFM Probe Calibration

SPM, in general, and SFM, in particular, are relatively new by the standards of the more 'traditional' techniques for nanoscale analysis. Thus, even though generically derived assumptions and information about the characteristics and properties of probes, such as those provided by a supplier, can be sufficient for most investigations, there is a growing trend towards the requirement of a particular probe to be calibrated before, after, and even during, an experiment in order to demonstrate 'quality assurance'. Failure to do so, and ignorance of the limitations inherent in the image formation process(es), could invalidate what might otherwise be a carefully designed and executed study. The topic of probe calibration is extensive, and cannot be covered in detail in a broad description of SPM techniques and applications. Nevertheless, significant issues are dealt with below, with an emphasis on SFM probes, and mainly by references to the literature.

5.2.3.3.1   Normal Spring Constant

In many cases the adoption of the manufacturer's nominal values for k_N will suffice, with due recognition of uncertainties. However, the actual value for a particular probe may be needed. The range of available procedures is summarized in Table 5.3 in the appendix to this chapter.

5.2.3.3.2 Lateral Spring Constant

The lever responds to in-plane force components, leading to torsional or longitudinal (buckling) bending, as well as to out-of-plane components. In the case of a single-beam diving-board configuration, and if the long axis of the lever is perpendicular to the fast scan direction, then only the torsional mode will be stimulated, to a good first-order approximation. If the torsional spring constant, and the relationship between the torsional angle at the tip and the PSPD response are known, then lateral forces can be quantified. Therein is the principle of LFM, (also known as 'chemical force microscopy'). In the so-called multi-asperity regime, in which friction is independent of contact area, the lateral force component, F_L, due to surface chemistry, can be written as

$$F_L = \mu * (F_N + SI_A) \tag{5.2}$$

where $\mu*$ is an equivalent coefficient of friction, F_N is the normal force, S is the area of contact, and I_A is the force of adhesion per unit area. The normal force F_N, is equal to $k_N \Delta z$, with k_N in the range $10^{-2} - 10^2$ N/m. Since the z-resolution of a typical SFM is better than 0.1 nm, the minimum detectable normal force is of order 10^{-12} N. Assuming a maximum compliance of the lever of 1 μm, the maximum normal force imposed by the lever will be 10^{-4} N. The corresponding approximate angles of deflection, for a lever of length 100 μm, will be 10^{-6} rad (resolution limit) and 10^{-2} rad (maximum compliance). The lateral force is also given by $F_L = k_L \Delta x$, where k_L is the lateral spring constant related to the torsional spring constant and the length of the tip, and Δx is the displacement of the apex of the tip perpendicular to the long axis of the lever. The angle of torsional rotation will be $\Delta x/h$ where h is the length of the tip. Given a limit on resolution of 10^{-6} rad, and a tip length of 5 μm, the minimum detectable Δx will be ca. 5 pm. Disregarding the effects of adhesion, the effective coefficient of friction is then

$$\mu* = \frac{F_L}{F_N} = \frac{k_L}{k_N} \times \frac{\Delta x}{\Delta z} \tag{5.3}$$

Values of $\mu*$ down to 10^{-3} can be measured with standard probes. Some manufacturers provide an estimate of the lateral spring constant. Analytical expressions have been derived for k_L for V-shaped levers (Neumeister and Ducker 1994). However, finite element analyses (Labardi et al. 1994) suggest that the bending modes of such levers can be more complex than those that can be modelled analytically. Beam-shaped levers in the long thin beam approximation have much simpler modes, and can be modelled with reasonable accuracy by the simple expression

$$k_L = \frac{Et^3 w}{6L(1+v)h^2} \tag{5.4}$$

where t, w, L are the thickness, width, and length of the lever, respectively, E is Young's modulus, ν is Poisson's ratio, and h is the length of the tip. The so-called wedge calibration provides a direct experimental method for determining k_L (Ogletree, Carpick, and Salmeron 1996; Carpick, Ogletree, and Salmeron 1997). A lateral electrical nanobalance for determination of lateral force constants has been described (Cumpson, Hedley, and Clifford 2005).

5.2.3.3.3   Tip Parameters

Aspect ratio, radius of curvature at the apex, and tip length are the three most critical geometrical parameters of the probe tip. The parameters affect the quality of images, particularly in the contact mode, and are the principal sources of artefacts (Atamny and Baiker 1995; Glasbey et al. 1994; Odin et al. 1994, and references therein). All three can be determined most conveniently by a method known as 'reverse imaging'. The principle is simple. If an actual tip is used to image a feature that is known to approximate to a delta-function spike, then the image of the spike will be a true representation of the tip shape (Montelius, Tegenfeldt, and van Heeren 1994). The manufacturers of SPM instrumentation can offer lithographically patterned Si surfaces with arrays of spikey features. An example of the application of reverse imaging is shown in Figure 5.9.

5.3   STM/STS

5.3.1   PHYSICAL PRINCIPLES—BRIEF THEORY

Electron tunnelling is the elementary process that underlies the operation of STM/STS. The approach adopted by Tersoff and Hamann 1985, remains an intuitive and useful tool for interpretation of data. Additional material can be found in the review literature (van Leemput and van Kempen 1992; Hofer 2003). The energy level scheme of a tunnel barrier is shown in Figure 5.10. The respective wave functions outside and inside the barrier are sinusoidal and exponential; the latter is particularly relevant to the problem at hand. The tunnel current is given by a summation over elastic tunnelling channels, (hence the presence of $\delta(E_2 - E_1)$ in the expression for the tunnel current, where E refers to energy as a variable, and the subscripts are defined below), linking occupied states (as described by the Fermi–Dirac function, f) on one side of the barrier, with unoccupied states $(1 - f)$ on the other side, as shown in equation (5.5).

$$I = \frac{2\pi e}{h} \sum_{1,2} f(E_2)[1 - f(E_1 + eV_T)] |M_{1,2}^2| \delta(E_2 - E_1) \tag{5.5}$$

The distributions of states in the tip and sample are displaced by eV_T, where V_T is the tunnel voltage. The diagram illustrates the case of tunnelling between two metal surfaces in close proximity, from occupied states below E_F (in metal 1) to unoccupied states above E_F (in metal 2) (where E_F is the Fermi energy). The matrix element $M_{1,2}$

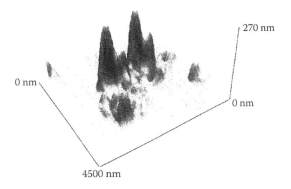

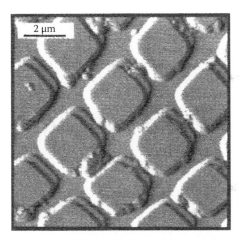

FIGURE 5.9 A double tip is revealed by recording reverse images of two carbon whiskers (top). The resultant spatially correlated artefact, an apparent ledge, for a semiconductor grid is shown (lower). (From Watson, G. S. and Myhra, S. unpublished data.)

is the transfer-Hamiltonian of Bardeen 1961 and is a weakly varying function for constant V_T.

The simple one-dimensional case of two identical free-electron metals with identical work functions, Φ, provides useful insight. Applying boundary conditions, the wave functions in the gap may be written as

$$\psi_1 = \psi_1^0 e^{-kz}$$
$$\psi_2 = \psi_2^0 e^{-k(s-z)}$$

(5.6)

where the width of the tunnel barrier is s; k is a real number (typically 10 nm^{-1}), and is given by $k = (2m\Phi/h^2)^{1/2}$. It is straightforward to show, since M is a weakly varying function, and if $eV_T < \Phi$ and $eV_T < E_F$, that

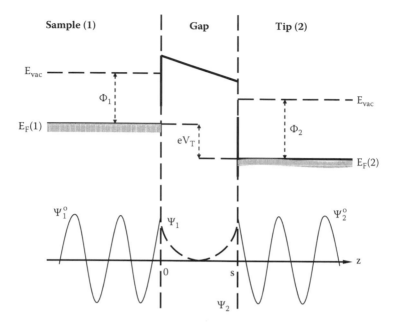

FIGURE 5.10 Energy level diagram and depiction of wave functions for a tunnel barrier. (From Myhra, S., 2010, in *Handbook of Surface and Interface Analysis*, Rivière, J. C. and Myhra, S., eds., with permission from Taylor & Francis.)

$$I \propto \sum_{1,2} \left|\psi_1^o\right|^2 \left|\psi_2^o\right|^2 e^{-2ks} \tag{5.7}$$

The expression gives qualitative insight into the tunnelling process, irrespective of tip shape, if the wave function anchored in the tip can be approximated by an s-wave. Furthermore, if the overlaps of the respective wave functions within the barrier are small, and if the tails of the two functions are of similar shape, then

$$I \propto \sum_{i} \left|\Psi_i\right|^2 \delta\left(E_1 - E_F\right) \tag{5.8}$$

It can now be recalled from solid-state physics that the local density of states in the sample surface at the position of the tip can be written as

$$\rho(E_F, r) = \sum_{1} \left|\psi_1(r)\right|^2 \delta(E_1 - E_F) \tag{5.9}$$

with *r* being a spatial coordinate in direct space. This description allows a number of qualitative statements about the STM/STS system to be made, as set out below. More sophisticated approaches to the interpretation of STM images have been reviewed in the literature (Hofer 2003).

The exponential dependence of the tunnel current I on the width of the tunnel gap accounts for the extreme spatial resolution in the z-direction; a change in s of 0.1 nm will change I by a factor of e^2.

The model shows that an STM image, formed by mapping constant tunnel current, actually represents a map of constant density of states (more correctly constant transition probability). Thus, unless there is a one-to-one spatial correlation between the positions of atoms, that is, nuclei, and maxima/minima in the density of states map, then it is the electronic structure that is being revealed. Conversely, the symmetries in the image must reflect the surface structural symmetries, if there is a one-to-one correspondence.

The lateral resolution in the x-y plane is due, in part, to the exponential dependence of I on the barrier width. As well, the lateral decay of the s-state located in the apex atom of the tip will confine the interaction laterally to a radius of ca. 0.1 nm. Thus, there will be single atom resolution due to the lateral spatial decay of the density of states function.

Of equal significance is the fact that the elastic tunnel process picks out particular states in the sample and tip, as a consequence of energy conservation. Thus, the tunnel voltage, V_T, sets an energy window, with a width of a few k_BT. Consequently, a value of the tunnel voltage can be chosen for maximum resolution, that is, an energy at which the corresponding density of electron states has the greatest spatial variation. These deductions account for the relative ease with which 'atomic' resolution can be obtained for covalent materials (e.g., Si); this is a consequence of the strong spatial dependence of the density of states across the unit cell at the extrema of bands. Conversely, similar resolution is much more difficult to obtain for metals where the spatial variations are more gentle.

The ability to 'tune' the energy window of the tunnelling process by changing eV_T is the basis for scanning tunnelling spectroscopy (STS). Successive maps of the surface at different values of V_T will reveal the real-space locations of the corresponding equi-energy contours of the density of states. Local spectroscopy can be carried out by fixing the lateral and vertical positions of the apex of the tip with respect to the surface and recording an I_T versus V_T curve. The I_T versus V_T data will not be interpretable as a simple plot of the density of states function, however, since there will be an exponential dependence of the tunnel probability with V_T from the matrix element M. This and other effects can be eliminated, in part, by measuring (dI/dV)/(I/V) which is a useful dimensionless quantity related to the local band structure. As well as being local, STS has the additional merits of extreme surface specificity—literally the first monolayer—and of being able to probe both valence and conduction band states by the simple expediency of reversing the polarity of V_T (and thus reversing the tunnel process). The pioneering work by Wolkow and Avouris (1988) is an excellent example.

From the above description of the tunnel process, it can be seen that the tunnel current is exponentially dependent on k, as well as on barrier width. The decay constant, k, is a function of the effective barrier height, and is therefore related to the work function of the surface being probed. Thus, it is possible to extract the relative spatial variation of the work function from the STM map.

It will also be apparent that STM/STS can be applied only to materials that are tolerably good conductors. The criterion for 'goodness' is that the effective sample resistance, R_S, must be such that $I_T R_S \ll V_T$. Hence, the technique will work well for clean surfaces of metals, alloys, semimetals, and doped semiconductors. The presence of thin (< 1 nm) insulating surface barriers (e.g., oxide layers) can be accommodated by allowing them to act as tunnel barriers. Because of the extreme surface specificity of STM/STS, an UHV environment is necessary when the surface is reactive. One non-UHV area of application in which the STM has much to offer is that of a fluid environment. In particular, electrochemical STM, where the tip can be a local electrode as well as a local probe, has turned out to be a useful area of application (e.g., Moffat 2007).

5.3.2   OPERATIONAL MODES—STM

It is useful to distinguish between the various techniques that are members of the SPM family (e.g., STM, AFM, etc.), and the operational modes of particular techniques (e.g., constant current versus constant height imaging in the case of STM; constant force versus constant height imaging in the case of AFM).

5.3.2.1   Imaging at Constant Tunnel Current

The feedback loop (see Figure 5.4) senses the difference signal between the tunnel current and a set-point; the loop then adjusts the z-height of the scanner in order to minimise the difference signal. The topographical map is derived from the relative z-stage excursion at each pixel location in the x-y field of view. The dynamic range (the 'depth of focus') is set by the limits of extension/contraction of the z-stage of the scanner. The mode is appropriate for relatively rough surfaces, and for large fields of view.

5.3.2.2   Imaging at Constant Height

The z-stage is essentially disabled by increasing the time constant of the feedback loop, and the z-height information is derived from the change in tunnel current at each pixel location. The mode is preferred for flat surfaces (cleaved crystal faces, epitaxial films, etc.,) and allows rapid scanning over small fields, thus minimising the effects of thermal drift.

5.3.2.3   Error Signal Mapping

The error signal mode is a variation on the constant height mode. In this case, the difference signal arising from the inability of the feedback loop to respond to sudden changes in slope, with the loop enabled, is sensed and plotted. Changes in the slope of the topography, for example, edges, will appear more prominently in the image, while more gentle changes in topographic height will be correspondingly washed out.

5.3.2.4   I-V Spectroscopy

The tip is positioned at a particular location in the x-y plane. The feedback loop is momentarily disabled while V_T is swept over the range of interest. The effect is to

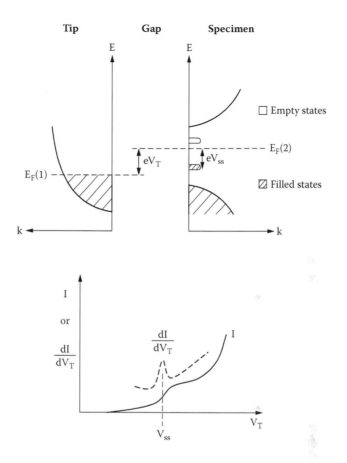

FIGURE 5.11 Schematic illustration of STM I-V spectroscopy (eV_{ss} is the energy of a surface state with respect to the Fermi energy). (From Myhra, S., 2010, in *Handbook of Surface and Interface Analysis*, Rivière, J. C. and Myhra, S., eds., with permission from Taylor & Francis.)

shift the Fermi edge of the tip with respect to energy states in the surface at the location of the tip. While the tunnel current is exponentially dependent on the voltage, additional contributions to the current will occur when the Fermi edge coincides with either the energy of a surface state, or a band edge, in the electronic structure in the surface. The result is shown schematically in Figure 5.11. The effect can be enhanced by plotting dI/dV_T, or $d(\ln I)/dV_T$) versus tunnel voltage.

5.4 SFM

A description of an SFM system cannot be carried out with the same degree of generality, neatness, and unity as for STM/STS. Neither can the image formation process be understood at the same level of detail and physical insight. Continuum theories will provide useful guidance, but phenomenologically, only a rough description of the system can be provided, which may, however, be adequate for most purposes.

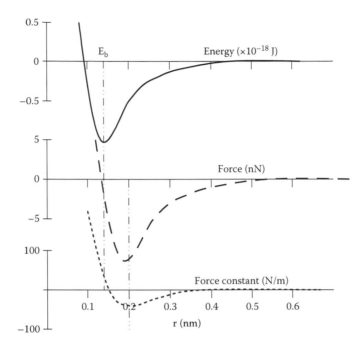

FIGURE 5.12 (Top) The attractive and repulsive forces acting on the tip combine to produce a potential well. (Middle) The resultant net force as a function of tip-to-surface distance. (Bottom) A plot of force constants, revealing high constants in the repulsive regime ($E < E_b$), and lower constants in the attractive regime ($E > E_b$). The force arising from the deflection of the lever maintains a quasi-static equilibrium. (From Myhra, S., 2010, in *Handbook of Surface and Interface Analysis*, Rivière, J. C. and Myhra, S., eds., with permission from Taylor & Francis.)

5.4.1 Physical Principles

It is illustrative to consider some of the general characteristics of interatomic interactions, as shown schematically in Figure 5.12. The short-range repulsive interaction, due to the non-classical exchange force, combines with the longer-range attractive ones, due to dispersion forces of the van der Waals type, to produce a potential well. There will be a location of lowest energy, the binding energy E_b, where the attractive and repulsive force components are balanced. Due to the shape of the potential well, the force constants in the repulsive part of the well are much greater than those in the attractive part. The latter observation is highly relevant to the design and characteristics of SPM instrumentation. In the repulsive contact mode, where the force constant is high, of order 100 N/m, the demands on spatial control in the z-direction are relatively modest, and good performance can obtained in the DC control mode. In the attractive non-contact mode, on the other hand, the force constant is low, of order 10 N/m. In practice, it becomes necessary to control the z-direction spatial position in the AC mode.

A Lennard–Jones type of potential function is usually adopted as a workable approximation for the contact mode. However, a full description would require extensive modelling (such as a full simulation by molecular dynamics (Israelachvili 1992)).

TABLE 5.1

Tip-To-Surface Interactions (D << R)

Interaction	Geometry	Expression for Force*
Capacitance	Sphere-to-flat	$\dfrac{\varepsilon\varepsilon_0\pi RV^2}{D}$
	Cone-to-flat	$\varepsilon\varepsilon_0\pi V^2 \tan^2\theta \ln\left(\dfrac{1}{D}\right)$
Charge versus fixed dipole	Charge-to-flat	$\pm\left(\dfrac{\pi\rho\mu}{2\pi\varepsilon\varepsilon_0}\right)\ln\left(\dfrac{D+t}{D}\right)$
Charge versus 'free' dipole	Charge-to-flat	$-\dfrac{\pi\rho q^2\mu^2}{3(\pi\varepsilon\varepsilon_0)^2 kT}\left(\dfrac{1}{2D^2}-\dfrac{1}{(D+t)^2}\right)$
Charge-induced dipole	Charge-to-flat	$-\dfrac{\pi\rho q^2 a}{3(\pi\varepsilon\varepsilon_0)^2}\left(\dfrac{1}{2D^2}-\dfrac{1}{(D+t)^2}\right)$
Capillary	Sphere-to-flat	$-4\pi R\gamma_{LV}\cos\theta + 4\pi R\gamma_{SL}$
van der Waals	Sphere-to-flat	$-\dfrac{HR}{6D^2}$
	Cone-to-flat	$-\dfrac{H\tan^2\theta}{6D}$
Fixed dipoles	Sphere-to-flat	$\pm\beta((D+2t)\ln(D+2t)+D\ln D-2(D+t)\ln(D+t))$
	Cone-to-flat	$\pm\eta((D+2t)^2\ln(D+2t)+D^2\ln D-2(D+t)^2\ln(D+t))$
Patch charges	Sphere-to-flat	$-\dfrac{\delta}{(D+A)^2}+\dfrac{\xi}{(2D+A+B)^2}$

Source: Adapted from Burnham, et al., 1993, *Nanotechnology.*

* R = radius of curvature of tip; V = applied potential; D = tip-to-surface separation; a = length of dipole; q = point charge; θ = half-angle of cone, or contact angle of meniscus; $\varepsilon_0\varepsilon$ = permittivity of free space and relative permittivity, respectively; ρ = charge density; μ = dipole moment; t = layer thickness; β = polarizability; $\gamma_{LV,SL}$ = surface energy (tension); LV,SL = liquid-to-vapour and solid-to-vapour, respectively; H = Hamaker's constant; ξ, β, η, δ = constants (see Burnham et al. 1993); A = location of charge on tip; B = radius of curvature of tip (for patch charge model).

In the non-contact mode, the van der Waals interaction plays a major role, but other types of interaction may need to be considered. Relevant geometries and useful expressions are summarised in Table 5.1 (adapted from Burnham, Colton, and Pollock 1993).

5.4.2 SFM OPERATIONAL MODES

5.4.2.1 AC Modes—Non-contact and Intermittent Contact (Tapping)

The probe is stimulated to oscillate at resonance as a free-running oscillator (well away from the surface). The shape of the resonance peak of the oscillating probe is

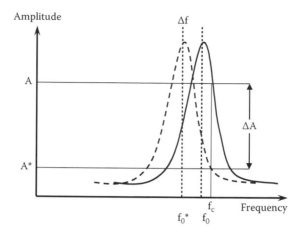

FIGURE 5.13 The principle of AC control for non-contact and intermittent contact imaging. A is the set-point amplitude at which the system is maintained, while ΔA is the (exaggerated) error signal. (From Myhra, S., 2010, in *Handbook of Surface and Interface Analysis*, Rivière, J. C. and Myhra, S., eds., with permission from Taylor & Francis.)

shown in Figure 5.13. When the tip enters the force field of the surface, the system is similar to that of a classical damped and driven anharmonic oscillator, described by the following equation.

$$\frac{d^2z}{dt^2} + \frac{\gamma dz}{dt} + \omega_0^2 z = \frac{A_0}{m*} \exp(i\omega t) \tag{5.10}$$

where $\gamma = b/m*$, with $m*$ the effective mass and b the drag coefficient, $\omega_0 = [k/m*]^{1/2}$ is the free-running resonance frequency with k the spring constant, A_0 is the amplitude of the driving force and ω is the driving frequency.

The system can then be described completely for various conditions (Sarid 1991). However, it is sufficient for a qualitative description to note that the resonance frequency of a free oscillator is

$$f_0 = \frac{1}{2\pi}\left[\frac{k_N}{m*}\right]^{1/2} \tag{5.11}$$

where k_N is the 'normal' spring constant of the lever. During the closest approach to the surface, the tip will sense an attractive force with an effective force constant $k*$. The shift in the characteristic frequency of the lever is then given by

$$f_0^* = \frac{1}{2\pi}\left[\frac{k_N - k*}{m*}\right]^{1/2} \tag{5.12}$$

The situation is shown schematically in Figure 5.13. The dependence of the characteristic frequency on the strength of tip-to-surface interaction can be exploited in mapping the topography of the surface. There are three methods of control and detection.

5.4.2.1.1   Slope Detection (Also Known as 'Amplitude Detection')

The probe is driven at a frequency f_c resulting in an amplitude A, below that of the free-running resonance amplitude. The conditions constitute the set-point. An increase, or decrease, in strength of tip-to-surface interaction, will give rise to a frequency shift, and thus a further decrement, or increment, in amplitude, to A*. The decrement/increment is then fed back to the control-loop, which generates an error signal. The error signal adjusts the z-height of the scanner, thus restoring the set-point conditions.

5.4.2.1.2   Phase Detection—Topographic Imaging

There is rapid change in the phase of the oscillating lever, with respect to the phase of the driving signal, at the peak of the resonance envelope, in response to topographic excursions in the z-direction. Thus, the system can be phase-locked and controlled at a fixed phase change; the fixed phase will then constitute the set-point for the feedback loop, while the phase increment/decrement constitutes the control signal being fed back to the loop. Most current instruments have adopted amplitude (slope) detection as the preferred method for topographical imaging in the intermittent contact mode.

5.4.2.1.3   Phase Detection—Mapping of Surface Stiffness

In this mode, the feedback loop is controlled by the amplitude/slope detection method. The damping term determines the phase increment (or decrement) of the oscillating lever with respect to the driving signal, and the damping is a function of the stiffness of the surface (i.e., more or less energy is being transferred from the oscillator to the surface). Thus, lateral variations in surface stiffness can be sensed and mapped with the aid of a phase-locked loop. This way of mapping is particularly useful for a two-phase surface, where one phase has a lower elastic modulus than the other phase (e.g., a two-phase polymer, or a composite consisting of ceramic particles embedded in a polymer matrix). An example is shown in Figure 5.14. It must be borne in mind that the topographic signal and the phase signal are not completely decoupled. Thus, the method is most reliable for phase mapping when the surface is relatively flat.

5.4.2.1.4   FM Detection

The tip-to-surface interaction sensed during the closest approach by the tip in the AC mode gives rise to a frequency shift, Δf, of the oscillating probe with respect to the fixed frequency of excitation. The shift increases with increasing strength of interaction. Thus, the feedback loop can be controlled by a signal proportional to the deviation from a set-point that is determined by a given frequency shift while maintaining constant amplitude of oscillation. In this manner, a surface map, based

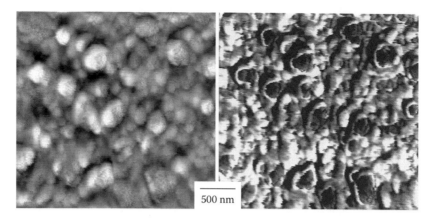

500 nm

FIGURE 5.14 Topographic (left) and phase contrast (right) AFM images of a surface of a polymer with $BaTiO_3$ nanoparticles embedded. The dark contrast features in the phase image represent hard objects (corresponding to a lesser phase shift than that of the softer polymer matrix). (From Zhao, 2008.)

on constant strength of interaction, can be generated. The FM AC mode may be preferable to slope detection for high Q resonance modes, that is, in vacuum, and for relatively flat surfaces.

5.4.2.2 LFM and Friction Loop Analysis

In-plane as well as out-of-plane force components will act on the tip at the point of contact with the surface. The former will exert a friction-like force on the tip in the direction of travel. If the fast-scan raster direction is perpendicular to the long axis of the lever, then the lateral friction force will cause a torsional deformation of the lever that can be sensed by the signal on the left-right PSPD segments (see Figure 5.8). Detection of lateral forces can be used to generate an image based on 'friction' contrast. Quantitative investigations of local friction require that the friction loop be analysed (Gibson, Watson, and Myrha 1997). The latter is illustrated in Figure 5.15. The torsional force constant of a lever, in the long and thin beam approximation can be written as

$$k_T = k_N \left[\frac{2L^2}{3(1+\mu)h^2} \right] \tag{5.13}$$

where the form of the expression shows that k_T, and thus the sensitivity to in-plane forces, depends on the normal force constant k_N, the length of the lever L, the length of the tip h, and Poisson's ratio, μ.

To a first approximation, the lateral force effect is similar to that of macroscopic friction between two objects in sliding contact (the lateral force is independent of relative speed, and independent of the 'contact' area). The relationship between lateral force, F_F, being sensed, and normal force, F_N, being imposed by the lever is given by

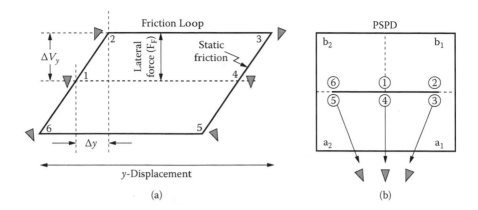

FIGURE 5.15 Schematic representation (a) of the friction loop. The fast scan direction is along the y-axis with the long axis of the lever being perpendicular to the fast scan direction. The response of the lever, (i.e., torsional deflection due to the in-plane 'friction' force), to lateral forces acting on the tip, is shown along the vertical axis (in exaggerated form). The lateral 'friction' force F_F can be quantified from the signal ΔV_y, while the displacement Δy is due to static 'friction'. The corresponding positions of reflected light incident on the PSPD are illustrated in (b). The Top-Bottom differential signal, sensed by the control loop, is the output $(b_1+b_2)-(a_1+a_2)$ from the arithmetic unit, and the Left-Right differential lateral signal is the output $(b_2+a_2)-(b_1+a_1)$. (From Myhra, S., 2010, in *Handbook of Surface and Interface Analysis*, Rivière, J. C. and Myhra, S., eds., with permission from Taylor & Francis.)

$$F_F = \mu^*(F_N + F_A) \qquad (5.14)$$

where μ^* is the coefficient of friction and F_A is the force of adhesion between tip and surface. AFM in the lateral force mode is now a routine tool for investigations of nanotribology and nanomechanics, (see review literature, e.g., Bhushan 2005). In the single asperity regime, in which a sharp tip is sliding across a hard surface, the relationship is more complex (Carpick and Salmeron 1997). A surface may be laterally differentiated by virtue of variation in surface chemistry (the differentiation may not reveal itself in the topographical contrast). Chemical contrast will result from chemical differentiation that manifests itself as a change in adhesive force and/or a change in in-plane force components. Hence, the LFM mode is often called 'chemical force microscopy' or 'friction force microscopy' (FFM) (Noy et al. 1995). The LFM imaging mode is particularly useful for delineating phase-separated polymer surfaces, Figure 5.16 (Li 2004) and for investigating the dependence of adhesion and friction on interface modification (Tsukruk and Blizniuk 1998).

5.4.2.3 F-d Analysis

Force-versus-distance analysis has been gaining in importance and popularity in comparison with imaging modes. F-d curves can provide information on local materials properties, such as elasticity, hardness, Hamaker constant, adhesion, and surface charge. F-d analysis has had a major impact on the understanding of fundamental

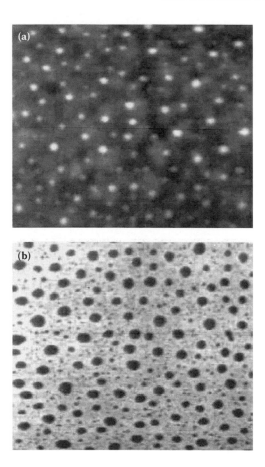

FIGURE 5.16 Topographic (A) and lateral force (B) images of a phase-separated polymer blend film (PMA/PMMA at 95:5 nominal weight fraction). The images show that the phase structure is more clearly delineated in the LFM than in the topographic imaging mode. (Adapted from Li, 2004, *Polymer Testing*, with permission from Elsevier.)

interactions in colloid science, and has shed new light on nanoscale friction and lubrication. A recent review covers F-d exhaustively and provides a near-encyclopaedic list of references (Butt, Capella, and Kappi 2005).

Quasi-static F-d analysis can be undertaken by holding the tip at a particular x-y location far away from the surface. The sample is then driven towards the tip at a rate that is slow in comparison with the mechanical response of the system. The net force is sensed during the approach, contact, and retraction parts of the cycle. An idealised response curve is shown in Figure 5.17, with stage travel and lever deflection plotted on the horizontal and vertical axes, respectively. The vertical scale can be converted into force sensed/applied by the lever by the simple expediency of multiplying the deflection by k_N. The plot in Figure 5.17 corresponds to the case when both tip and surface are incompressible, and when the surface is covered with a thin adsorbed aqueous film. The various segments represent:

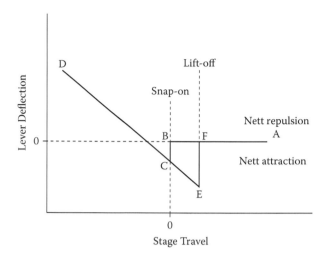

FIGURE 5.17 Idealised F-d curve for a probe interacting with a 'hard' surface. The system consists of an incompressible tip and a surface, with the surface being covered by an adsorbed aqueous film.

Approach half-cycle:
- **AB**—tip and surface are well-separated, no interaction.
- **BC**—the tip senses the attractive interaction from the meniscus layer, the force constant of interaction exceeds k_N, and the tip 'snaps' into contact with the 'hard' surface. There is a regime of instability due to the force constant of interaction being greater than that of the lever.
- **CD**—tip and surface are incompressible, and the stage travel distance must therefore be equal to the lever deflection.

Retract half-cycle:
- **DE**—the system retraces itself since all deformations/deflections are elastic.
- **EF**—the meniscus interaction has increased due to capillary action, and there will be greater vertical spacing at the lift-off instability/discontinuity (than for the snap-on).
- **FA**—return to large separation and no interaction.

This kind of plot is representative of events for an air-ambient instrument due to adsorbed moisture. The meniscus interaction is generally a nuisance feature in that it will mask other surface mechanical effects. The meniscus can be eliminated by carrying out F-d analysis under water (or some other fluid ambient). Alternatively, in the case of an instrument operated within a vacuum envelope, the aqueous phase will be pumped away. However, neither the snap-on or the lift-off discontinuities can be eliminated entirely, due to the force constant of interaction being greater than k_N during the approach half-cycle, and there will always be an adhesive tip-to-surface contact interaction that must be overcome during the lift-off half-cycle.

The 'hard' surface F-d curve can be used to calibrate the detection system so that a measurable detector response is related accurately to the lever deflection. Applications of F-d analysis in most areas of science and technology are now ubiquitous. The impact on the study of the nanomechanical properties of soft materials (e.g., organics, polymers, and biomaterials) has been particularly significant.

Surfaces and structures, which are compliant with force constants comparable to that of the lever, will give rise to F-d responses that are different from those shown in Figure 5.17. A 'rogues' gallery of curves is shown in Figure 5.18. The case in which

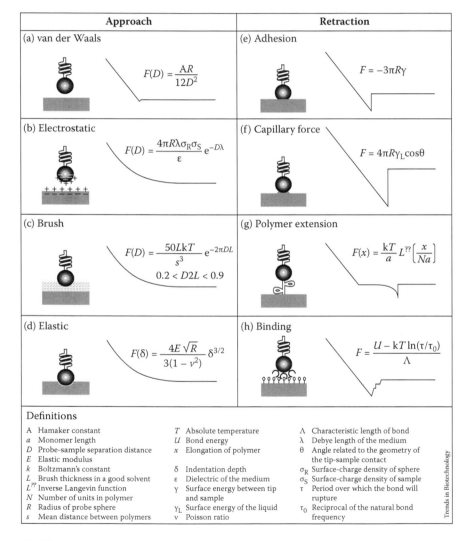

Approach	Retraction
(a) van der Waals $F(D) = \dfrac{AR}{12D^2}$	(e) Adhesion $F = -3\pi R\gamma$
(b) Electrostatic $F(D) = \dfrac{4\pi R\lambda\sigma_R\sigma_S}{\varepsilon} e^{-D\lambda}$	(f) Capillary force $F = 4\pi R\gamma_L\cos\theta$
(c) Brush $F(D) = \dfrac{50LkT}{s^3} e^{-2\pi DL}$ $0.2 < D2L < 0.9$	(g) Polymer extension $F(x) = \dfrac{kT}{a} L^{??}\left[\dfrac{x}{Na}\right]$
(d) Elastic $F(\delta) = \dfrac{4E\sqrt{R}}{3(1-v^2)} \delta^{3/2}$	(h) Binding $F = \dfrac{U - kT\ln(\tau/\tau_0)}{\Lambda}$

Definitions

A Hamaker constant	T Absolute temperature	Λ Characteristic length of bond
a Monomer length	U Bond energy	λ Debye length of the medium
D Probe-sample separation distance	x Elongation of polymer	θ Angle related to the geometry of
E Elastic modulus		the tip-sample contact
k Boltzmann's constant	δ Indentation depth	σ_R Surface-charge density of sphere
L Brush thickness in a good solvent	ε Dielectric of the medium	σ_S Surface-charge density of sample
$L^{??}$ Inverse Langevin function	γ Surface energy between tip	τ Period over which the bond will
N Number of units in polymer	and sample	rupture
R Radius of probe sphere	γ_L Surface energy of the liquid	τ_0 Reciprocal of the natural bond
s Mean distance between polymers	v Poisson ratio	frequency

Trends in Biotechnology

FIGURE 5.18 Some examples of F-d curves and the expressions that describe the dependence of force on tip-to-surface distance. (a)–(d) represent the approach cycle and (e)–(h) represent the retract cycle. (From Heinz and Hoh, 1999, *Trends in Biotech.*, with permission from Elsevier.)

the tip is indenting a compliant surface is particularly useful. The AFM is then effectively configured as a nanoindenter, where the lever and the sample become two compliant elements in series. Due to the finite aspect ratio of the tip, the approach curve will usually be non-linear, with a slope lower than that of the calibration curve (shown in Figure 5.17). The stage travel is then equal to the sum of lever deflection and surface indentation. Since the force constant of the lever, k_N, and the respective compressions of the two compliant elements are known, the stiffness of the surface can be calculated as a function of applied force. The non-linearity of the indentation curve, illustrated in Figure 5.18d, can be deconvoluted in order to infer an equivalent tip-shape (Blach et al. 2001), using standard expressions for the relationships between applied force, depth of indentation, and tip shape, with Young's modulus as a parameter (Snedden 1965). The procedure allows calculation of the elastic modulus. Additional information can be obtained in the F-d mode when the AFM is acting as a nanoindenter:

- The snap-on and lift-off forces, at the points of discontinuity in the approach and retract curves, can be measured at the points of greatest net attraction. The latter is generally taken to be the force of adhesion (Figure 5.18e). However, due to the uncertainty in the area of contact, the force of adhesion cannot reliably be converted into a force per unit area. Likewise, the surface chemistry of the tip is generally not known with any degree of certainty, and detracts further from estimates of adhesion.
- The extent of hysteresis in the system is a measure of the work of indentation, and is given by the area enclosed by the approach and retract curves.

Typical F-d curves for different systems, and associated expressions describing the dependence of force on probe-to-sample separation, are shown in Figure 5.18, from Heinz and Hoh (1999).

A consequence of the discussion above is the need to match the force constant of the lever to that of the interaction being investigated in order to extract maximum information. If there is mismatch, then either the lever will be the only compliant element, and no information is obtained about the surface, or the surface will be the only compliant element, and the deflection of the lever is not measurable.

5.4.2.3.1   Stiffness and Young's Modulus of Soft Materials

Measurements of the stiffness of materials, and subsequent calculation of a Young's modulus, are traditionally carried out by indentation with a hard tip of known geometry (e.g., a diamond stylus with a triangular conical Berkowitz shape). An indentation measurement typically has a macroscopic interaction volume, although some instruments can reduce this to the near-μm range, and quantification is dependent on being able to measure the width and depth of an indentation pit (e.g., Larsson et al. 1996).

An AFM operated in the F-d mode as a nanoindenter has the additional advantage of providing a nanoresolved topographic image. Such an image is shown in Figure 5.19, while the F-d results from that and similar surfaces are displayed in Figure 5.20.

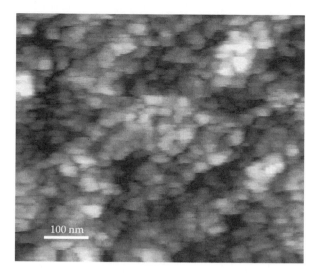

FIGURE 5.19 Tapping mode topographic AFM image of an aerogel surface, consisting of a network of particles in the 5–20 nm size range. (From Myhra, 2010, unpublished results.)

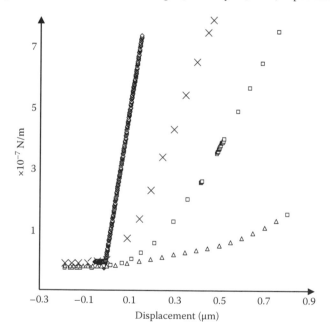

FIGURE 5.20 Force versus distance curves for three aerogel formulations of different densities. The calibration curve, that with the steepest slope, was recorded from a Si wafer. The three curves with successively lower slopes represent increasing indentation of the samples for the same force loading; they correspond to decreasing density, and thus, to an increasingly dilute network consisting of chains of silica particles. The nominal force constant for the lever was 4.5 N/m. The nominal radius of the apex of the tip was 10 nm, and the opening half-angle of the cone-shaped tip was 10°. (From Myhra, 2010, unpublished data.)

TABLE 5.2

Results from Nanoindentation of Aerogel Samples of Different Densities

Batch	Indentation (nm)	Sample Stiffness k_s (N/m)
1	90	1.1
2	250	0.4
3	650	0.15

The depths of indentation at constant force and the stiffness for the three formulations are listed in Table 5.2.

Nanoindentation in the AFM F-d mode has also been applied to nanostructured biological materials; an example is shown in Figure 5.21.

5.4.2.3.2 *Colloidal Probe Analysis*

This method of carrying out F-d analysis was initially adopted for the measurement of colloidal forces in an aqueous fluid arising from surface charges (e.g., in Ducker, Senden, and Pashley 1991). A growing number of AFM-based studies of biomolecular interactions have been carried out by sensing forces in the sub-nN range, versus distance (i.e., standard F-d analysis), between a Si or Si_3N_4 tip prepared with a particular (bio)chemical functionality and a functionalized surface (in order to investigate the interaction between antibody and antigen molecular species, where the species have been attached to the tip and surface, respectively). More often, however, a bead (of diameter from less than 1 to several μm) is attached to the tip, so as to produce a 'colloidal' probe, an arrangement that has advantages for investigations of intermediate- and short-range interactions (Watson et al. 2003). The geometry of the probe bead is then known with greater relative certainty than that of the tip. Likewise, the surface chemistry of a bead can be prepared with greater flexibility and reliability than in the case of a bare tip; a wide range of well-characterized microspheres/beads with known surface (bio)chemistries can be obtained from several suppliers. Also, the much greater surface area will effectively amplify the strength of interaction and thus improve the signal-to-noise ratio. On the other hand, there will be greater uncertainty as to the number of interacting species, and lateral spatial resolution is substantially degraded, in comparison with a sharp probe tip. In the case of a functionalised tip, with a radius of curvature less than 50 nm, the number of interacting species may be less than ten, thus allowing single-event binding forces to be estimated from a histogram of the data. A reliable method for attachment at the tip apex of a single nanoparticle, of diameter in the range 10–50 nm, with known surface chemistry, has been reported (Vakarelski and Higashitani 2006), as shown in Figure 5.22.

5.4.2.3.3 *Biomolecular Binding and Unfolding*

The F-d operational mode is now used for nanomechanical analysis throughout science and technology. An early description (Burnham and Colton 1989) provides a

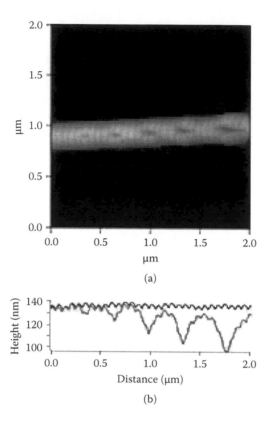

(a)

(b)

FIGURE 5.21 (**See color insert.**) (a) Plastic deformation of a collagen fibril resulting from indentation by AFM for loads up to 10 μN. (b) Contour line along the centre of the fibril in (a) shows indentation pits of increasing depth and width. The black line corresponds to the un-deformed fibril, with the corrugations showing the periodicity of the so-called D-band spacing. The image was obtained in the tapping mode. (Adapted from Grant et al., 2009, *Biophys. J.*, with permission from Elsevier.)

good introduction to the basic ideas. The merits of F-d are particularly apparent in studies of biomolecular binding (Willemsen et al. 2000; Zlatanova, Lindsay, and Leuba 2000, and references therein) and protein (un)folding. An example of the former is shown in Figure 5.23, where both the location and the strength of a single antibody-antigen recognition event have been demonstrated (Hinterdorfer et al. 1996).

Every long-chain molecule has its own unique folded native structure, which will depend on the environment and on the stage of the biomolecule in its functional life-cycle. Due to the enormous number of degrees of freedom, it is a daunting task to calculate its minimum energy configuration. Nevertheless, a polypeptide chain will find its folded global minimum energy configuration in a remarkably short time (Levinthal's paradox (Levinthal 1968)). Explanations currently favoured revolve around the existence of an identifiable directed pathway through the multi-dimensional potential landscape. One possible approach to the gaining of insight into the problem is to reverse-engineer the folding process, namely to induce unfolding.

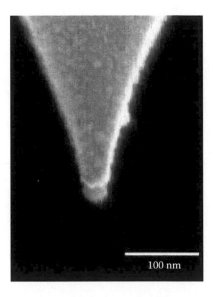

FIGURE 5.22 SEM image of a single gold nanoparticle with a diameter of ca. 25 nm positioned at the apex of a standard silicon nitride probe tip. (Adapted from Vakarelski and Higashitani, 2006, *Langmuir*, with permission from the American Chemical Society.)

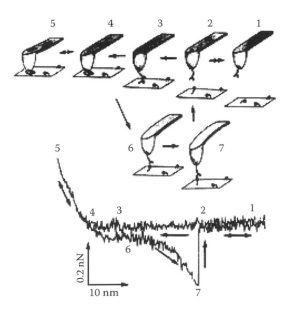

FIGURE 5.23 The schematic sequence shows (1–5) approach of the antibody and attachment to the antigen, followed by (6) retraction of the probe and stretching of the tether and bond. Finally, the antibody-antigen bond is ruptured at (7) due to the lever-imposed force of ca. 200 pN. Actual data in the form of an F-d curve are shown below. (From Hinterdorfer et al., 1996, *Proc. Natl. Acad. Sci.*, with permission from National Academy of Sciences, USA.)

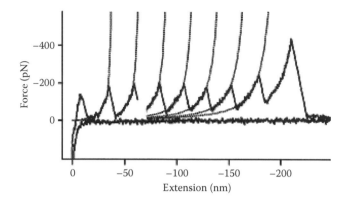

FIGURE 5.24 A force versus extension graph of an octameric TI 127 poly-protein being pulled by an AFM probe. The upper curve represents a sequence of continuous extension curves (rising portions of the trace) and sudden un-folding events. The lower trace represents the initial approach. (From Best and Clarke, 2002, *Chem. Commun.*, with permission from the Royal Society of Chemistry.)

It has been shown that an AFM operated in the F-d mode can shed considerable light on the problem (Best and Clarke 2002). A folded protein is attached to a substrate without being denatured (i.e., without undergoing change from its 'natural' condition). The probe tip, often prepared with a particular chemical surface functionality, is first attached to a reactive site, through trial and error, and then withdrawn slowly, while a force is applied, causing extension initially, and then subsequently leading to a sequence of unfolding stages. A typical outcome is shown in Figure 5.24.

5.5 SCM

5.5.1 Principles and Implementation

Scanning capacitance microscopy is a relatively recent and specialised technique, but is gaining in importance and popularity for semiconductor device characterization in response to the shrinking dimensions of such devices. Under favourable conditions, surface conductivity can be mapped with a lateral resolution of 5 nm (Álvarez et al. 2003). Its principal merits are those of being non-destructive and of having lateral spatial resolution in the 5–10 nm range. As shown in Figure 5.25 the configuration of tip and specimen becomes that of a metal-oxide-semiconductor (MOS) capacitor. The equivalent circuit has a resonance frequency, ca. 1 Ghz, determined by the impedance of the transmission line (arising from a capacitance of a few pF and an inductance of a few nH). If a frequency shift of 1 khz can be detected, then, in principle, the sensitivity is of the order of 10^{-18} F.

5.5.2 Capacitance Mapping

The method for mapping lateral variations in capacitance is illustrated in Figure 5.26. In essence, a control point is established at a frequency corresponding to the steepest

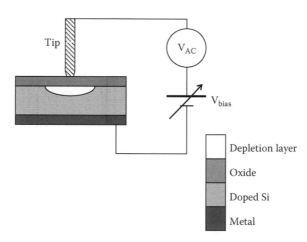

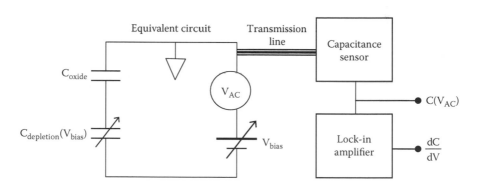

FIGURE 5.25 In SCM the tip/specimen system can be modelled as an MOS device in which the metallic tip constitutes the gate electrode. Extent of depletion below the gate dielectric layer is controlled by V_{bias}. The equivalent circuit consists of a grounded tip in series with two capacitors also in series. In combination with the transmission line, the system will have a resonant frequency of ca. 1 Ghz. (From Myhra, S., 2010, in *Handbook of Surface and Interface Analysis*, Rivière, J. C. and Myhra, S., eds., with permission from Taylor & Francis.)

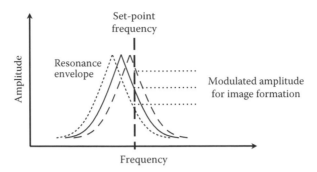

FIGURE 5.26   The resonance envelope shifts in response to change in local capacitance, giving rise to a corresponding change in amplitude at the set-point frequency. (From Myhra, S., 2010, in *Handbook of Surface and Interface Analysis*, Rivière, J. C. and Myhra, S., eds., with permission from Taylor & Francis.)

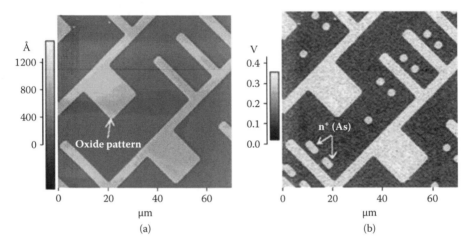

FIGURE 5.27   (**See color insert.**) Contact mode topographic (left), and SCM (right), images of a patterned semiconductor surface. Regions with bright contrast in the topographic image represent a thermally grown SiO_2 pattern of height 70 nm. The additional circular and rounded rectangular features evident in the SCM image correspond to regions heavily doped by ion implantation (50 keV As^+ ions and 10^{14} ions/cm^2 dose density). (From Yoo et al., Park Systems Web site, SCM Mode Note, www.parkafm.co.kr. With permission.)

part of the resonance envelope. As the local capacitance changes, the resonance envelope will shift, and thus give rise to a difference in output voltage for a fixed set-point frequency. Examples are shown in Figures 5.27 and 5.28.

5.5.3   MAPPING DIFFERENTIAL CAPACITANCE

The applied AC signal, in combination with the DC bias, causes a time-dependent variation in the extent of depletion, and thus, a change in local capacitance. If the corresponding change in amplitude, or phase, is detected by a lock-in amplifier, at

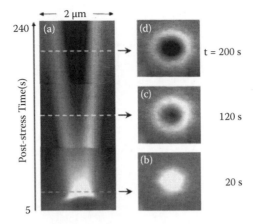

FIGURE 5.28 Charge redistribution mapped as a function of time. The charge was trapped initially on the surface of a 6nm thick silicon oxide on p-doped Si by a voltage stress (−8 V applied by the tip). The SCM images were obtained with the instrument controlled in the contact topographic mode, while an AC modulation signal of 50 mV at 50 kHz was applied to the tip. The fields of view in (b)–(d) are 2×2 μm², and the bright/dark contrast refers to net positive/negative charge. (From Mang et al., 2004, *Europhys. Lett.*, with permission from EPS.)

the AC set-point frequency, then the differential capacitance, dC/dV, can be mapped. The effect of bias voltage on capacitance and differential capacitance is illustrated in Figure 5.29.

5.6 SNOM

It is well known that the spatial resolution of optical imaging in the far-field region is limited by diffraction to slightly less than half the wavelength of the incident light. The possibility of imaging in the near-field regime was recognised more than 70 years ago (Synge 1928), but was given serious consideration as a practical proposition only in the early 80s with the advent of SPM technologies (Pohl and Nowotny 1994; Kirstein 1999; Hecht et al. 2000).

5.6.1 PHYSICAL PRINCIPLES

While SFM techniques have many unique attributes, they cannot provide analytical or vibrational spectroscopic information. As well, the current generation of SFM instrumentation has relatively poor temporal resolution. SNOM extends the spatial resolution of classical optical microscopy by a factor of ten or more, in combination with modest spectroscopic capability and excellent temporal resolution. SNOM exploits the properties of evanescent fields, which can be confined on structures much smaller than the wavelength of the incident light. Evanescent fields do not propagate into the far-field region; thus, local probe techniques can be adapted either to illumination of the sample or to the detection of the evanescent fields emitted by the sample. Commercial instruments are based on the former principle, whereby the

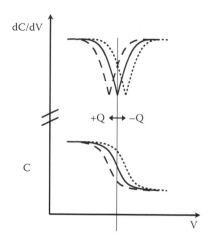

FIGURE 5.29 The effect on capacitance and differential capacitance of local accumulation or depletion of charge as a function of applied voltage. Addition or subtraction of charge causes the characteristic curves to shift left or right, respectively. (From Myhra, S., 2010, in *Handbook of Surface and Interface Analysis*, Rivière, J. C. and Myhra, S., eds., with permission from Taylor & Francis.)

field is emitted from, and defined by, an aperture (or a metallic tip) of dimensions ca. $\lambda/20$.

An object illuminated by an external field will become polarised, and will re-emit some of the absorbed radiation. The scattered radiation fields have evanescent components that are bound to the scatterer, and non-evanescent components that can propagate into the far field. Conversely, polarisation of matter will also occur when an external field interacts with an extremely small volume of matter. In the latter case, the evanescent field in the vicinity of the scatterer will contain information about the geometry of the scatterer.

The currently favoured arrangement is that of the generation of an optical near-field at a small aperture at the apex of a tapered fibre. If a sample is located within a few nm of the aperture, then the sample will be submerged in, and be coupled to, near-field radiation. The sample will act as a scatterer of near-field radiation, resulting in information being available in the far field, in transmission or reflection modes. It is important to recognise that the light injected into the fibre will be subject to interactions with the whole system—that is, during transmission along the fibre, by the aperture, and from coupling to the sample. Thus, detailed modelling must take account of the whole system. Much work has been done on theoretical investigations of the transmission coefficient of the fibre. It turns out that the main limitations are the size of the aperture, and the angle of the tapered cone. Coefficients of transmission in the range 10^{-3}–10^{-6} are required for practical exploitation. Calculation of the actual field distribution at the aperture, when a sample is present, is non-trivial, even though it can be carried out within the framework of classical electrodynamics. The state-of-the-art has been described in several excellent reviews (e.g., Girard, Joachim, and Gauthier 2000; Girard 2005).

5.6.2 Technical Details

A schematic of a SNOM instrument is shown in Figure 5.30. It is usually based on an inverted microscope platform where the existing conventional optics are adapted to collect transmitted and reflected intensity in the far-field region. Likewise, conventional optical components can be introduced into the optical paths in order to filter, disperse, or polarize scattered light. Other components that are specific to the SNOM technique, and which affect its performance, need further description.

5.6.2.1 Shear-Force Detection

Unlike the more widely used SPM techniques, the 'strength' of interaction—for example, tunnel current in the case of STM and interatomic forces in the case of AFM —cannot easily be used to control the z-position of the SNOM probe. Instead, shear-force detection has been adopted as the most widely used method. The fibre tip is stimulated to oscillate laterally with an amplitude of 1–5 nm at its fundamental resonance by a tuning fork arrangement driven by an exciting signal at the requisite frequency. In the vicinity of the surface, shear forces will cause damping, resulting in an amplitude decrement and a change in phase. Several mechanisms can be the source of damping, for example, interaction with the adsorbed moisture film, intermittent contact with the surface, and/or electrostatic image forces. Normally, the phase change is monitored and constitutes the control signal that allows the feedback loop to maintain the probe at a constant distance from the surface. Thus, an SNOM can function in an AFM imaging mode, albeit at a somewhat lower z-resolution.

5.6.2.2 Optical Fibre Probe

The final probe is the outcome of a two-step processing of a standard optical fibre. The first step consists of preparing a transparent tapered conical section with a sharp apex. Two methods are used, either local heating and pulling (Betzig et al. 1991), or chemical etching in hydrofluoric acid (HF) (the so-called Turner's technique) (Turner 1984; Hoffmann, Dutoit, and Salathé 1995). The former is relatively labour-intensive and lacks reproducibility, but leaves a smooth surface finish. The latter is more convenient, lending itself to higher through-put parallel fabrication, but may not result in an optimum surface finish. The second step consists of deposition of a metal coating and creation of an aperture at the apex. Angle-resolved vapour deposition can be used to form a self-aligned aperture at the apex. More recently, focused ion-beam micromachining has been used to produce reproducible apertures of high quality (Pilevar et al. 1998; Veerman et al. 1998). A typical outcome is shown in Figure 5.31.

5.6.3 Operational Modes

The most widely used modes include transmission imaging and fluorescence detection. Imaging in the reflection mode can also be carried out, although at lower signal-to-noise ratio due to the smaller angle of detection and the need to work with an objective lens at lower numerical aperture. In either case, features in the optical

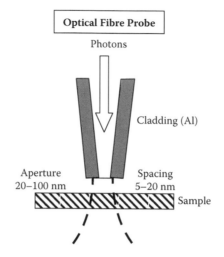

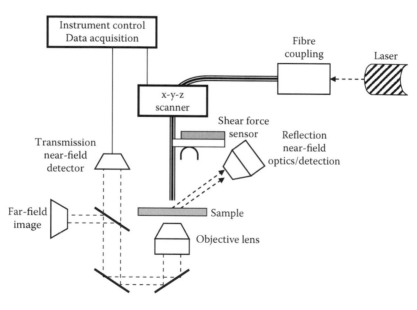

FIGURE 5.30 Schematic of an SNOM instrument. The optics for detection of near-field radiation in transmission or reflection are based on a standard inverted microscope platform. Polarisation contrast can be implemented by insertion of polarizing elements in the optical path before and after scattering by the specimen. Spectroscopic information in transmission can be obtained by inserting a dispersive component or a notch filter into the far-field optical path. (From Myhra, S., 2010, in *Handbook of Surface and Interface Analysis*, Rivière, J. C. and Myhra, S., eds., with permission from Taylor & Francis.)

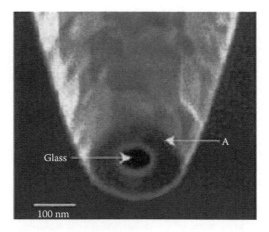

FIGURE 5.31 Aperture obtained by focused ion beam (FIB) milling. (From Veerman et al., 1998, *J. Microsc.*, with permission from Wiley.)

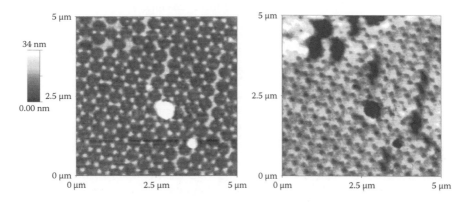

FIGURE 5.32 **(See color insert.)** Shear force and optical transmission images, left and right, respectively, of a standard SNOM specimen. The specimen was produced by metal evaporation onto a flat substrate with close-packed polystyrene spheres as a shadow-mask (the spheres were subsequently removed). (With kind permission from D. Higgins, 2007, Kansas State University.)

image can be correlated with those in a shear-force topographical image during parallel imaging (see Figure 5.32).

The most recent addition is that of near-field Raman spectroscopy, where in practice SERS enhancement is required in order to compensate for the low incident power (typically less than 100 nW).

5.6.3.1 Transmission Imaging

5.6.3.2 Fluorescence

Fluorescence analysis is particularly useful for locating labelled compounds within larger bio/organic structures. An example is shown in Figure 5.33. The technique is gaining in popularity.

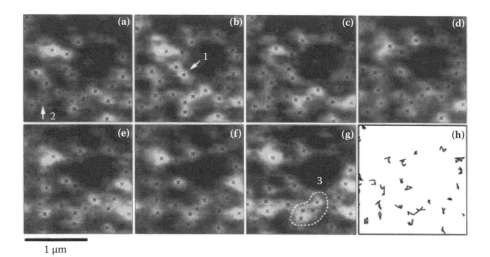

1 µm

FIGURE 5.33 Consecutive SNOM fluorescence images, obtained at 13 min intervals, of single molecules (Rhodamine-6G) embedded in polyvinylbutyral. The sequence A–G illustrates lateral mobility of individual molecules at ambient conditions. The resultant trajectories are shown in H. (From Bopp et al., 1996, *Chem. Phys. Lett.*, with kind permission from Elsevier.)

5.6.3.3 Near-Field Raman Spectroscopy and Mapping

Raman spectroscopy is of great value as a tool for structural finger-printing (see Chapters 7, 9, and 11). In the microprobe version, conventional far-field Raman mapping can resolve and identify structures with spatial resolution of ca. 1 µm. However, the 'normal' Raman process is relatively inefficient; only 1 in 10^7 incident photons will undergo the characteristic scattering. SERS enhancement can improve the yield by several orders of magnitude, thus making that form of Raman spectroscopy a practical proposition as an SNOM operational mode. The technique is currently being developed, and can be implemented with existing instrumentation. Several examples have been described in the literature (e.g., Girard 2005). As well, the dependence on polarisation can be investigated in a fluid environment (Grausem, Humbert, and Burneau 1997).

5.6.4 Tip-Enhanced Raman Spectroscopy (TERS)

Recent advances in aperture-less near-field optical microscopy, often referred to as TERS, have reached the stage at which instruments are available commercially.

A full description of the underlying mechanisms is beyond the scope of the present volume. However, it can be stated that the metallic tip can act as a nanoscopic scatterer (effectively acting as a nanoantenna), as well as defining the confinement of the electromagnetic field in the proximity of a nanoscale specimen. In the latter instance, the tip will enhance locally the scattered radiation due to the coupling with free electrons in the tip, thus stimulating the surface plasmons. An additional contribution comes from the geometric enhancement of the electromagnetic field around a sharp tip. Metals such as Cu, Ag, and Au, and tip radii in the range 5–50 nm, have been reported as being particularly effective. In general, all the effects tend to be present to some degree in the generation of the evanescent near-field. Images and spectral information recorded in the far-field are thus a result of tip-induced scattering of the evanescent field and tip-enhanced transmission of the near-field (see Kharintsev et al. 2007, and references therein).

In the case of Raman scattering, SERS can contribute additional enhancement, if the specimen is placed on a rough noble metal substrate. The enhancement is then due to a combination of effects such as interaction between the local surface plasmons and the dynamic plasmons, and chemical effects arising from transfer of energy into the vibrational modes of adsorbed species.

Due to the localisation of the enhanced near-field at the apex of the tip, the spatial resolution of spectroscopic information can reach 10 nm in favourable cases (Maultzsch, Reich, and Thomsen 2002).

5.6.4.1 Technical Implementation of TERS

Although the first generation of SNOM instruments was implemented on a dedicated platform, the TERS technique can be considered as an appendage to a general SPM instrument platform. Either a metallic STM probe, or a standard coated AFM probe, with focussed laser illumination incident on the tip apex, can act as the source of near-field illumination of the specimen, as opposed to the fibre tip aperture probe of the early generation SNOM. Consequently, the cost of the probe will be no greater, in the case of TERS analysis, than that for AFM/STM. A second major advantage is that the topographical imaging quality will be equivalent to that of having an AFM/STM operational mode available, as opposed to the degraded quality available with an SNOM probe and its tuning fork feedback control arrangement.

Three TERS configurations suitable for the analysis of opaque specimens are illustrated in Figure 5.34. In each case, spectral information is collected in the back-scatter mode, while the probe is submerged in the incident excitation and/or back-scattered radiation containing the spectral information. Efficiency of excitation and collection can be enhanced by the use of ×100 objective lenses and high numerical apertures. The scattered radiation is transported into the far-field where it is dispersed by standard optical spectrometry backed by high-quantum-yield parallel detection. If the sample is optically transparent, then TERS can be implemented on a standard inverted microscope base, with detection of scattered radiation in the transmission mode.

The performance of SNOM in the TERS configuration is illustrated in Figure 5.35.

The effect of SERS enhancement in combination with TERS is illustrated in Figure 5.36.

5.7 SECM

SECM is an extension of conventional spreading current analysis with the additional advantages of greater lateral spatial resolution, < 5 nm in favourable cases, and the ability to correlate I-V measurements with features in a high-resolution topographic map.

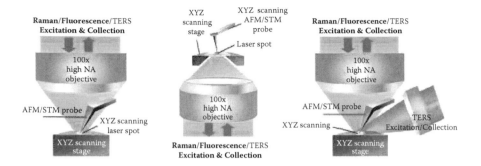

FIGURE 5.34 (**See color insert.**) Possible TERS configurations. (Left) A high numerical aperture objective brings the incident radiation to a focus at the tip apex. The diffraction-limited resolution of the objective is ca. 400 nm in air. The configuration is appropriate for AFM imaging and spectral analysis of opaque samples. (Middle) This configuration is based on an inverted microscope base, and is appropriate for spectral analysis in transmission with a nanoscale specimen deposited on a microscope slide. (Right) Here the illumination is incident from the side, with scattered radiation collected in the backscatter mode. However, the arrangement can also be reversed, with vertically incident illumination, and scattered radiation collected from the side (at the expense of a smaller solid angle of collection). (Adapted from Dorozhkin et al., 2010, *Microsc. Today,* November, p. 28.)

5.7.1 Physical Principles

The technique is based on contact mode AFM, with a sharp conducting tip making a point contact with a surface at a particular location in the x-y plane. The lateral extent of the point contact will depend on the shape of the tip at the apex, the normal force loading, and the stiffness of the surface. The technique can function as a straight spreading resistance probe, if the circuit is ohmic, or as a tunnelling probe, if there are one or more tunnel junctions in the circuit. Likewise, local defect states can be imaged as functions of applied bias. (The method will then essentially be similar to STS analysis, but with the difference that the states are buried. Thus, the quality of the analysis will not depend on an UHV environment.) High-sensitivity current sensing is then required, typically in the fA range, in order to obtain tunnelling I-V data for thin dielectric films. The current, due to either direct or Fowler-Nordheim tunnelling, will depend on the thickness and dielectric strength of the film, leakage paths, and charge traps (De Wolf, Brazel, and Erikson 2001).

5.7.2 Technical Details and Applications

In the case of an ohmic circuit and with the tip in contact with a semi-infinite homogeneous solid, the resistance of the solid can be inferred from the simple expression $R = \rho/2d$, where R is the measured resistance (from V/I), ρ is the resistivity, and d is the diameter of the (circular) contact junction. The underlying assumption is that the probe-to-surface contact resistance is negligible. A variety of materials have been used in order to ensure ohmic contact, durability of the probe, and the prevention of the formation of rectifying and/or insulating interface layers; they include highly doped Si, diamond, Pt/Ir, Co/Ir, Au, and so forth. Also, it is important that the

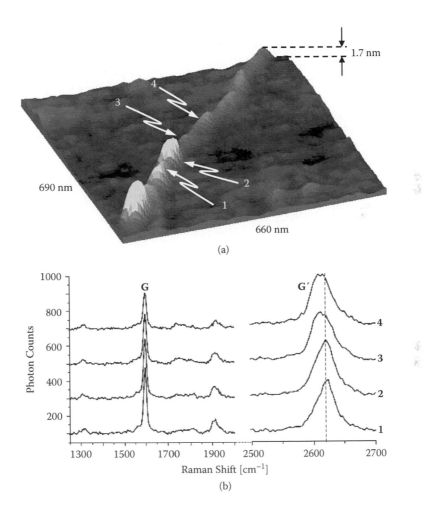

(a)

(b)

FIGURE 5.35 **(See color insert.)** (Top) Topographic image, shown in 3-D representation, of a SWCNT, grown by arc discharge, on a glass substrate. The apparently excessive width of the CNT was due to tip shape convolution. (Bottom) Near-field Raman spectra recorded at positions 1 to 4 in the image. The spacing between positions 2 and 3 was ca. 35 nm. The spectra are offset in the interest of clarity. (Adapted from Hartschuh et al., 2003, *Phys. Rev. Letts.*)

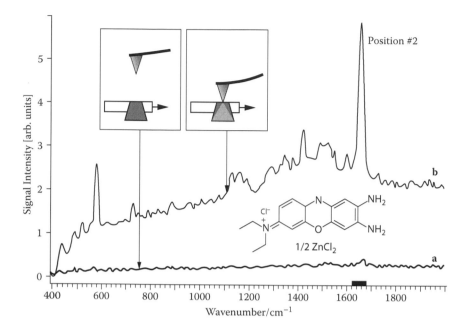

FIGURE 5.36 Raman analysis by TERS/SNOM in the transmission mode, illustrating the merit of surface enhancement (SERS). The AFM probe is silver-coated and illuminated by 488 nm radiation at 5 mW incident power. The sample consisted of a thin layer of BSB (brilliant cresyl blue with formula repeat unit shown as an inset) on a transparent substrate. The lower and upper traces represent data collection with the probe lifted off, and in contact with the surface, respectively. The SERS enhancement, when in contact, was ca. 2000, at an estimated lateral spatial resolution comparable to the size of the probe, ca. 50 nm. (From Stöckle et al., 2000, *Chem. Phys. Lett.*, with permission from Elsevier.)

measurements be undertaken on a flat, < 0.5 nm RMS roughness, surface, in order to ensure that features in a current map are not obscured by variations in contact area.

The most widespread application of the technique is in the area of characterization of electronic devices and materials: for example, a recent study of an SOI (silicon on insulator) device (Álvarez et al. 2003). A number of case studies have been discussed by De Wolf, Brazel, and Erikson (2001). A sequence of current images on a SiO_2 film (Figure 5.37) demonstrates the voltage-dependent leakage due to embedded defects.

The effect of tip conditions, in the context of an investigation of direct and Fowler-Nordheim tunnelling through insulating surface barriers on oxidised Si and a diamond-like carbon (DLC) film, has been demonstrated (Myhra and Watson 2005). The results are shown in Figure 5.38.

5.8   SCANNING KELVIN PROBE (SKP)

5.8.1   Effects of Electrostatic Interaction

There will be an electrostatic interaction between a surface charge distribution and a conductive tip; this can be modelled by an image charge calculation. If the charge

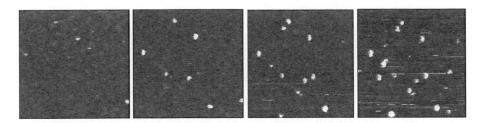

FIGURE 5.37   Tunnelling current images, revealing embedded defects in an SiO$_2$ film at increasing bias of 1, 2, 3, and 4V, from left to right. The field of view was 1×1 μm^2. (From De Wolf et al. 2001, *Mater. Sci. in Semicond. Processing,* with permission from Elsevier.)

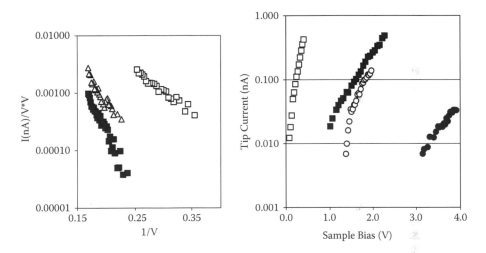

FIGURE 5.38   (Left) I-V data plotted to emphasise a fit with a Fowler–Nordheim tunnelling model. The logarithmic vertical axis refers to tip current in nA divided by the square of the sample bias. From left to right the sets of data refer to: as-received n-type tip with native oxide against n-type Si with 2.4 nm oxide; etched tip against Si with 2.4 nm oxide; and as-received tip against a diamond-like film (DLC). (Right) I-V characteristics for the following combinations (left to right): Au-coated tip against DLC; etched tip against DLC; oxidised tip against Au; oxidized tip against DLC. The semi-log plot emphasises the exponential dependence due to direct tunnelling. (From Myhra and Watson, 2005, *Appl. Phys. A*, with kind permission from Springer Science+Business Media.)

density is laterally differentiated, then the tip can be rastered across the surface, and a map of the electrostatic force can be produced. In the AC mode, where the tip is a damped and driven anharmonic oscillator, the change in force constant of interaction will manifest itself by changes in the resonance frequency, amplitude, or phase (with respect to the phase of the driving signal). The system will respond in a fashion similar to that discussed above under the heading of tapping mode imaging, or scanning capacitance imaging.

The electrostatic interaction is long range, (typically 100 nm), in comparison with interatomic forces (<10 nm for the attractive van der Waals exchange force and 0.1

nm for the repulsive force). Thus, in the contact or tapping imaging modes, either the shorter range forces will dominate, or there will be a mixing of interactions. One strategy, the so-called dual pass, is to produce a topographic image in the normal tapping mode, then withdraw the tip beyond 10 nm, and run an electrostatic force image (a similar strategy is generally adopted in magnetic force microscopy). This allows laterally differentiated surface charge densities to be correlated with the topographic map (albeit at the expense of limited spatial resolution for the charge map, and the effect of thermal drift).

An alternative, and more recent, strategy is to operate in a bimodal AC fashion (Magonov and Alexander 2010). The topographic image signal is generated from the normal sensing of the error signal by a lock-in amplifier with respect to the main resonance mode of the lever. In addition, a second oscillatory mode of the lever is stimulated, at a lower frequency, ω_{elec}, and lower amplitude, where the amplitudes of the two modes are additive (the signal that carries the topographic information is modulated by the frequency that carries information about the electrostatic interaction). The information carried by ω_{elec} modulation can be extracted from the overall signal by a second lock-in amplifier. Thus, the two images, topographic and electrostatic, can be acquired in parallel.

The surface potential can be measured by forcing the signal that arises from the electrostatic interaction to zero; this is performed by applying a DC bias to the probe, thus defining the local work function with respect to the work function of the tip. In this sense, the technique is a scanning Kelvin probe microscope.

Additional information can be obtained by monitoring the signal at $2\omega_{elec}$, which is proportional to the vertical gradient of tip-to-surface capacitance, dC/dz. The gradient is related to the dielectric permittivity.

An example of information that is available at the three frequencies is shown in Figure 5.39.

A more formal description of the SKP can be given, based on the discussion in a recent review (Melitz et al. 2011).

The contact potential difference (*CPD*) between the tip and sample is defined as

$$V_{CPD} = \frac{\phi_t - \phi_s}{-e} \tag{5.15}$$

When the tip is in close proximity to the sample surface, an electrical force is generated between the tip and sample, due to the difference in their respective Fermi energy levels. The situation is shown in Figure 5.40. In Figure 5.40a, the tip and sample surface are separated by a distance d and are not electrically connected (the vacuum levels are aligned, but the Fermi energy levels are different). On electrical connection (Figure 5.40b) the Fermi levels will be aligned, dipole layers are created, and a potential barrier, eV_{CPD}, arises due to the difference in work functions between tip and sample. (Note that V_{CPD} in Figure 5.40b should read eV_{CPD} in order that the units are consistent.) Thus, an electrostatic force will be acting on the probe. That force can be varied via an external reverse bias, V_{DC}; in Figure 5.40c, eV_{DC} is shown as set equal to the difference in the work functions.

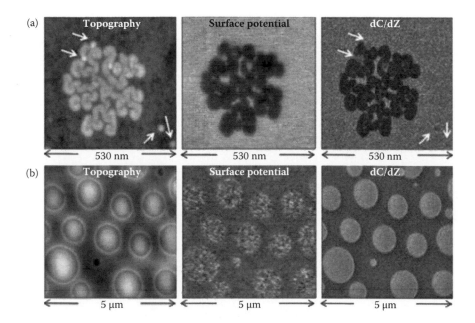

FIGURE 5.39 (See color insert.) Row A shows images of a fluoroalkane adsorbate $(F_{14}H_{20})$ on a Si substrate. The ranges of contrast are 0–7 nm for topography, 0–1.1 V for surface potential, and 0–0.15 V for dC/dz. Row B shows images for a 70 nm thick film of polystyrene/poly(vinyl acetate) 1:1 blend on an ITO (indium-tin oxide) substrate. The ranges of contrast are 0–28 nm for topography, 0–0.6 V for surface potential, and 0–0.11 V for dC/dz. (From Magonov and Alexander, 2010, *G.I.T. Imaging Microscopy*, with permission from S. Magonov.)

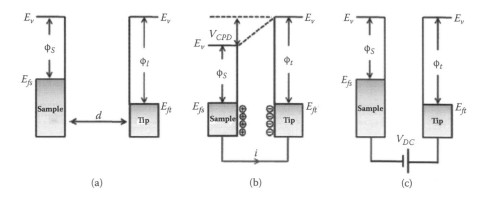

FIGURE 5.40 Electronic energy levels of the sample and AFM tip for three cases: (a) tip and sample separated by distance d with no electrical contact, (b) tip and sample in electrical contact, and (c) the application of an external bias (V_{DC}) between tip and sample to oppose the CPD. E_v is the vacuum energy level. E_{fs} and E_{ft} are the Fermi energy levels of sample and tip, respectively. (Adapted from Melitz et al., 2011, *Surf. Sci. Reports*, with permission from Elsevier.)

The electrostatic force, F_{es}, between tip and sample can be written as

$$F_{es}(z) = -\frac{1}{2}(\Delta V)^2 \frac{dC(z)}{dz} \qquad (5.16)$$

where ΔV is the potential difference ($V_{CPD} - V_{DC}$), and dC/dz is the gradient of the capacitance between tip and sample. There is an additional contribution of the carrier frequency, due to the oscillatory voltage $V_{AC} \sin \omega_{elect} t$ also applied to the tip. Therefore, the potential difference will be

$$\Delta V = (V_{DC} \pm V_{CPD}) + V_{AC} \sin \omega_{elect} t \qquad (5.17)$$

The force on the tip can therefore be written as

$$F_{es} = -\frac{1}{2}\frac{\partial C(z)}{\partial z}[(V_{DC} \pm V_{CPD}) + V_{AC} \sin \omega_{elect} t]^2 \qquad (5.18)$$

The three contributions to the force are then

$$F_{DC} = -\frac{1}{2}\frac{\partial C(z)}{\partial z}(V_{DC} \pm V_{CPD})^2$$

$$F_{\omega(elect)} = -\frac{\partial C(z)}{\partial z}(V_{DC} \pm V_{CPD})V_{AC} \sin \omega_{elect} t \qquad (5.19)$$

$$F_{2\omega(elect)} = \frac{1}{4}\frac{\partial C(z)}{\partial z}V_{AC}^2(\cos 2\omega_{elect} t - 1)$$

The term F_{DC} causes a static deflection of the tip, while V_{CPD} can be extracted from the expression for $F\omega_{(elect)}$. The term $F_2\omega_{(elect)}$ can be used for capacitance microscopy (see Section 5.5 in this chapter). The various parameters, that is, topographic z-height, V_{DC}, and dC/dz, can be acquired simultaneously at each (x,y) pixel location in a single pass raster, and then mapped as separate correlated images.

An atomically resolved *CPD* map for KBr is shown in Figure 5.41.

5.9 SCANNING ION CURRENT MICROSCOPY (SICM)

SICM can be considered as the ion current analogue of STM. A glass nanopipette is filled with an electrolyte, and a feedback loop maintains the pipette at a non-contact distance from the sample by controlling the ion current against a set-point. Imaging is carried out with the specimen immersed in a liquid buffer. The technique is particularly well suited for imaging of fragile and/or adhesive biological materials (Korchev et al. 2000; Pastré et al. 2001).

One electrode is located inside the pipette while the other is located in the solution sample cell. A bias is applied between the two electrodes, resulting in a current

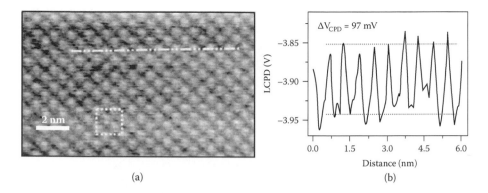

(a)   (b)

FIGURE 5.41   (a) A Kelvin probe force image of a single crystal KBr (001) surface recorded under UHV conditions. (b) Line trace of CPD (dot-dashed line in (a)); a potential difference of 97 mV is observed between K$^+$ and Br$^-$ sites. The dashed box in (a) indicates the unit cell. (Adapted from Bocquet et al., 2008, *Phys. Rev. B,* with kind permission from L. Nony.)

flow. The operating point is established by taking into account the resistance in the pipette channel and other probe-related variables. When the pipette is distant from the specimen, this resistance is at a fixed saturated value (assuming that the resistance in the fluid is negligible, so that the ionic mobility and the geometry of the capillary become the limiting factors, at constant bias). The approach of the probe to the specimen can then be controlled by the additional resistance, giving rise to an ion current below the saturation level, due to the increasing volume constraint, (a smaller effective cross-section for the conduction path), on the ionic conductance path, and thus a diminishing ionic source term. Imaging can be carried out in the DC constant current mode, or in the AC bias modulation mode. The lateral resolution can be enhanced by recording an image in the error mode (where the difference between set-point and measurement is plotted), see Figure 5.42.

5.10   FUTURE PROSPECTS

5.10.1   Increased Spatial Resolution for SFM

It has long been received wisdom that only UHV-STM can deliver true single-atom resolution. There are a few examples of genuine single atom resolution for contact mode AFM, but only for chemically inert surfaces under exacting conditions (Ohnesorge and Binnig 1993), while similar resolution has been reported for the imaging of more reactive surfaces, that is, Si(111) (7×7), in the small-amplitude non-contact mode (Giessibl 1995). Recently, great progress in UHV-SFM has been achieved (Giessibl 2005; Hembacher, Giessibl, and Mannhart 2004). Indeed, the adoption of very stiff levers (k_N >10^3 N/m), allowing non-contact operation at amplitudes of ca. 0.1 nm, has resulted in subatomic resolution. The information content in the image can be increased further by the detection of higher harmonics (at higher frequency). Comparative STM and AFM (higher harmonic detection) images are shown in Figure 5.43. While these techniques cannot be described as routine, they do offer directions for future improvements in instrumentation.

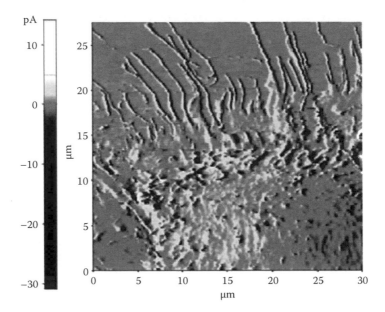

FIGURE 5.42 SICM current image recorded in the error mode of a live mouse muscle cell (type C2C12). The grey-scale colour scale bar is in pA. (From Schürmann, M. and Anselmetti, D. unpublished results, with permission.)

5.10.2 SINGLE ATOM CHEMICAL IDENTIFICATION BY AFM

Although the SPM family has provided a vast amount of information, it has so far been unable to provide direct compositional information. However, advances in small-amplitude intermittent contact (tapping) imaging, in which the combination of long-range van der Waals interaction and the short-range exchange force is being sensed, have allowed genuine single atom spatial resolution imaging by SFM techniques (see Section 5.10.1). Those techniques have been extended to the acquisition of F-d data over a range of less than 1 nm, and at a force resolution in the low pN range, in the regime where the atom at the apex of the tip is making an intermittent 'bond' with the closest atom in the surface (Hembacher, Giessibl, and Mannhart 2004). The technique is based on the shape of the F-d curve within 0.2 nm of the greatest force of attraction unique to a particular atom species, thus allowing differentiation of species. The map in Figure 5.44 illustrates the specificity of identification of atomic species in a field of view where simple imaging cannot discriminate any differences. The data were obtained in an UHV instrument at room temperature in the frequency modulation (FM) (see Section 5.4.2.1.4) control mode.

The above technique places considerable demands on spatial control and stability. In principle, a moderate vacuum should be sufficient to achieve the required conditions (i.e., a high-Q resonance mode and low thermal drift). However, in contrast to the ultra-high resolution mode described in Section 5.10.1, the data can be obtained at room temperature. Thus, there is considerable scope for implementing SFM as a true analytical technique.

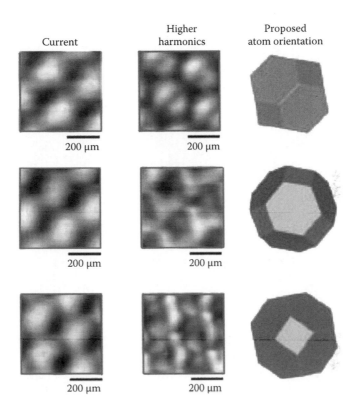

Current Higher harmonics Proposed atom orientation

200 μm 200 μm

200 μm 200 μm

200 μm 200 μm

FIGURE 5.43 Conventional ('current') constant-height UHV-STM (left column) and higher-harmonic UHV-AFM images (central column) of graphite, obtained with a W tip. The right-hand column shows the proposed orientation of the W tip atom which is represented by its Wigner–Seitz unit cell, reflecting the full symmetry of the bulk. The bonding symmetry of the protruding tip atom is assumed to be similar to the bonding symmetry of the bulk. Top row: the higher harmonic measurement shows a two-fold symmetry, resulting from an [110] orientation of the tip atom. Second row: using higher harmonics a three-fold symmetry is found, as expected for a [111] orientation. Third row: the symmetry of the higher-harmonic signal is approximately four-fold, corresponding to a tip in a [001] orientation. The imaging was carried out at 4.9 K with k_N = 1800 N/m, amplitude = 0.3 nm, resonance frequency = 18.0765 khz, and with Q = 20 000. (Adapted from Giessibl, 2005, *Materials Today*, with permission from Elsevier.)

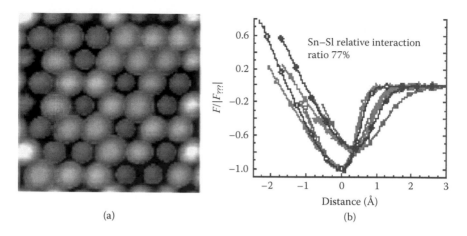

(a)   (b)

FIGURE 5.44 (**See color insert.**) (Left) False colour image of a surface alloy composed of Si (red-brown), Pb (green), and Sn (blue) atoms deposited on an Si(111) surface. The field of view is 4.3×4.3 nm². (Right) Normalized F-d curves showing the distinct difference in interaction between the tip and Si and Sn surface atoms (the corresponding results for Pb-Si and Si-Si interactions are not shown). Curves of different colour refer to data obtained with different probes, and they demonstrate that the assignments are independent of the tip material. The F-d data have been normalised to those recorded for the Si species ($F/F_{Si\text{-}set} = 1$ at distance = 0). (Adapted from Sugimoto et al., 2007, *Nature*, with permission from Macmillan Publishers Ltd.)

5.10.3 FASTER SCAN RATES

The principal reason for increasing scan rates is to track surface and interface dynamics in real time and thus increase the knowledge base for the thermodynamics and kinetics of processes. TV scan rates are likely to emerge in next-generation instruments when smaller probes with higher resonance frequencies become available. In addition, the mechanical stiffness of the instrumental closed-loop must increase correspondingly, in unison with faster electronics for control and data acquisition. Increased scan rates will have the added benefits of ensuring greater convenience in the laboratory and greater through-put in the factory.

5.10.4 GREATER INTEGRATION AND SPECIALISATION

There is likely to be a trend towards integrating SPM with complementary techniques, for example, LV-SEM (LV = low voltage), HRTEM (HR = high resolution), and surface analytical techniques. Conversely, while the multi-mode multi-technique modularity of SPM instrumentation will continue to lead the market, there will be increasing emphasis on producing SPM instrumentation that has been optimised for niche markets, for example, the electronic device industry and the life sciences.

APPENDIX: METHODS FOR CALIBRATION
OF NORMAL FORCE CONSTANT, K_N

Table 5.3 summarises the literature concerned with calibration of the normal force constant of an SFM force sensing/imposing probe.

TABLE 5.3
Methods for Determination of k_N

Method	References	Accuracy	Comments/Merits/Demerits
Dynamic Response Methods			
Resonance frequency with added mass	A	≈ 10%	Positioning and calibration of load difficult; potentially destructive
Thermal fluctuations	B	10%–20%	Temperature control essential; only suitable for soft levers; requires analysis of resonance curve
Simple scaling from resonance frequency	C	5%–10%	Depends on dimensional accuracy and determination of effective mass. Convenient, often implemented in instrument software
Theoretical Methods			
Finite difference method	D	>10%	Depends on dimensional accuracy and Young's modulus
Parallel beam approx.	E F	>10%	Depends on dimensional accuracy and Young's modulus
Static Response Methods			
Static deflection with added mass	E	15%	Positioning and calibration of load difficult; potentially destructive
Response to pendulum force	E	30%–40%	Complex and time-consuming procedure
Static deflection with external standard	G	15%–40%	Requires accurate external standard

Note: A: Cleveland, J. P., Manne, S., Bocek, D., and Hansma, P. K., 1993, *Rev. Sci. Instrum.*, 64, 403. B: Hutter, J. L. and Bechhoefer, J., 1993, *Rev. Sci. Instrum.,* 64, 1868. C: Sader, J. E., Larson, I., Mulvaney, P., and White, L. R., 1995, *Rev. Sci. Instrum.*, 66, 3789. D: Neumeister, J. M., and Ducker, W. A., 1994, *Rev. Sci. Instrum.*, 65, 2527. E: Senden, T. J. and Ducker, W. A., 1994, *Langmuir*, 10, 1003; Butt, H. J., Siedle, P., Seifert, K., Fendler, K., Seeger, T., Bamberg, E., Weisenhorn, A. L., Goldie, K., and Engel, A., 1992, *J. Microsc.*, 169, 75. F: Sader, J. E., 1995, *Rev. Sci. Instrum.*, 66, 4583. G: Gibson, C. T., Watson, G. S., and Myhra, S., 1996, *Nanotechnology*, 7, 259.

REFERENCES

Álvarez, D., Hartwich, J., Fouchier, M., Eyben, P., and Vandervorst, W., 2003, *Appl. Phys. Lett.*, 82, 1724.

Atamny, F. and Baiker, A. 1995, *Surf. Sci.*, 323, L314.

Bardeen, J. 1961, *Phys. Rev. Lett.*, 6, 57.

Best, R. B. and Clarke, J., 2002, *Chem. Commun.*, 183.

Betzig, E., Trautmann, J. K., Harris, T. D., Weiner, J. S., and Kostalek, R. L., 1991, *Science*, 251, 1468.

Bhushan, B., 2005, *Wear*, 259, 1507.

Binnig, G. and Rohrer, H., 1982, *Helv. Phys. Acta*, 55, 726.

Binnig, G., Quate, C. F., and Gerber, C., 1986, *Phys. Rev. Lett.*, 56, 930.

Blach, J. A., Watson, G. S., Busfield, W. K., and Myhra, S., 2001, *Polym. Intl.*, 51, 12.

Bocquet, F., Nony, L., Loppacher, C., and Glatzel, T., 2008, *Phys. Rev. B*, 78, 035410.

Bopp, M. A., Meixner, A. J., Tarrach, G., Zschokke-Gränacher, I., and Novotny, L., 1996, *Chem. Phys. Lett.*, 236, 721.

Burnham, N. A. and Colton, R. J., 1989, *J. Vac. Sci. Technol. B*, 7, 2906.

Burnham, N. A., Colton, R. J., and Pollock, H. M., 1993, *Nanotechnology*, 4, 64.

Butt, H. J., Capella, B., and Kappi, M., 2005, *Surface Science Reports*, 59, 1.

Carpick, R. W., Ogletree, D. F. and Salmeron, M., 1997, *Appl. Phys. Lett.,* 70, 1548.

Carpick, R. W. and Salmeron, M., 1997, *Chem. Rev.*, 97, 1163.

Cumpson, P. J., Hedley, J., and Clifford, C. A., 2005, *J. Vac. Sci. Technol. B*, 23, 1992.

De Wolf, P., Brazel, E., and Erickson, A., 2001, *Mater. Sci. in Semicond. Processing*, 4, 71.

DiNardo, N. J., 2007, *Nanoscale Characterization of Surfaces and Interfaces*, Wenheim: VCH.

Dorozhkin, P., Kuznetsov, E., Schokin, A., Timofeev, S., and Bykov, V., 2010, *Microsc. Today*, November, p. 28.

Ducker, W. A., Senden, T. J., and Pashley, R. M., 1991, *Nature*, 353, 239.

Gibson, C. T., Watson, G. S., and Myhra, S., 1997, *Wear*, 213, 72.

Giessibl, F. J., 1995, *Science*, 267, 68.

———, 2005, *Materials Today*, May.

Girard, C., 2005, *Rep. Prog. Phys.*, 68, 1883.

Girard, C., Joachim, C., and Gauthier, S., 2000, *Rep. Prog. Phys.*, 63, 893.

Glasbey, T. O., Batts, G. N., Davies, M. N., Jackson, D. E., Nicholas, E. V., Purbrick, M. D., Roberts, C. J., Tendler, S. J. B., and Williams, P. M., 1994, *Surf. Sci.*, 318, L1219.

Grant, C. A., Brockwell, D. J., Radford, S. E., and Thomson, N. H., 2009, *Biophys. J.*, 97, 2985.

Grausem, J., Humbert, B., and Burneau, A., 1997, *Appl. Phys. Lett.*, 70, 1671.

Hansma, P. K. and Tersoff, J., 1987, *J. Appl. Phys.,* 61, R1.

Hartschuh, A., Sanchez, E. J., Xie, X. S., and Novotny, L., 2003, *Phys. Rev. Letts.*, 90, 095503.

Hecht, B., Sick, B., Wild, U. P., Deckert, V., Zenobi, R., Martin, O. J. F., and Pohl, D. W., 2000, *J. Chem. Phys.*, 112, 7761.

Heinz, W. F. and Hoh, J. H., 1999, *Trends in Biotech.*, 17, 143.

Hellemans, L., Waeyaert, K., and Hennau, F., 1991, *J. Vac. Sci. Technol. B*, 9, 1309.

Hembacher, S., Giessibl, F. J., and Mannhart, J., 2004, *Science*, 305, 380.

Higgins, D., 2007. The images were kindly provided by Dr. D. Higgins, Kansas State University.

Hinterdorfer, P., Baumgartner, W., Gruber, H. J., Schilcher, K., and Schindler, H., 1996, *Proc. Natl. Acad. Sci. U.S.A*, 93, 3477.

Hofer, W. A., 2003, *Prog. Surf. Sci.*, 71, 147.

Hoffmann, P., Dutoit, B. and Salathé, R. P., 1995, *Ultramicroscopy,* 61, 165.

Israelachvili, J. N., 1992, *Intermolecular and Surface Forces*, 2nd ed., San Diego, CA: Academic Press.

Kharintsev, S. S, Hoffmann, G. G., Dorozhkin, P. S., deWith, G., and Loos, J., 2007, *Nanotechnology*, 315502.

Kirstein, S., 1999, *Current Opinion in Coll. Interface Sci.*, 4, 256.

Korchev, Y. E., Negulyaev, Y. A., Edwards, C. R. W., Vodyanoy, I., and Lab, M. J., 2000, *Nature Cell Biol.*, 2, 616.

Labardi, M., Allegrini, M., Solerna, M., Frediani, C., and Ascoli, C., 1994, *Appl. Phys. A*, 59, 3.

Larsson, P. L., Giannakopoulos, A. E., Söderlund, E., Rowcliffe, D. J., and Vestergaard, R., 1996, *Intl. J. Solids Struct.*, 33, 221.

Levinthal, C., 1997, *J. Chim. Phys.*, 65, 44.

Li, W.K., 2004, *Polymer Testing*, 23, 101.

Magonov, S. and Alexander, J., 2010, *G.I.T. Imaging Microscopy*, 4, p. 26.

Magonov, S. N. and Whangbo, M.H., 1996, *Surface Analysis with STM and AFM*, Wenheim: VCH.

Mang, K. M., Kuk, Y., Kwon, J., Kim, Y. S., Jeon, D., and Kang, C. J., 2004, *Europhys. Lett.*, 67, 261.

Maultzsch, J., Reich, S., and Thomsen, C., 2002, *Phys. Rev. B*, 65, 233402.

Melitz, W., Shen, J., Kummel, A. C., and Lee, S., 2011, *Surf. Sci. Reports*, 66, 1.

Meyer, E., Hug, H. J., and Bennewitz, R., 2004, *Scanning Probe Microscopy: The Lab on a Tip*, Berlin: Springer.

Moffat, T. P., 2007, *Electrochemical STM, in Encyclopedia of Electrochemistry*, Wiley-VCH.

Montelius, L., Tegenfeldt, J. O. and van Heeren, P., 1994, *J. Vac. Sci. Technol. B*, 12, 2222.

Myhra, S., 2010, In *Handbook of Surface and Interface Analysis*, Rivière, J. C. and Myhra, S., eds., Boca Raton: CRC.

Myhra, S. and Watson, G. S., 2005, *Appl. Phys. A*, 81, 487.

Neumeister, J. M. and Ducker, W. A., 1994, *Rev. Sci. Instrum.*, 65, 2527.

Noy, A, Frisbie, C. D., Rozsnyai, L. F., Wrighton, M. S. and Lieber, C. M., 1995, *J. Am. Chem. Soc.*, 117, 7943.

Odin, C., Aimé, J. P., El Kaakour, Z., and Bouhacina, T., 1994, *Surf. Sci.*, 317, 321.

Ogletree, D. F., Carpick, R. W., and Salmeron, M., 1996, *Rev. Sci. Instrum.*, 67, 3298.

Ohnesorge, F. and Binnig, G., 1993, *Science,* 260, 1451.

Scanning Capacitance Microscopy (SCM) Park Systems, South Korea, www.parkafm.co.kr.

Park Systems, Scanning Capacitance Microscopy (SCM), South Korea, www.parkafm.co.kr.

Pastré, D., Iwamoto, H., Liu, J., Szabo, G., and Shao, Z., 2001, *Ultramicroscopy*, 90, 13

Pilevar, S., Edinger, K., Atia, W., Smolianinov, J., and Davis, C., 1998, *Appl. Phys. Lett.*, 72, 3133.

Plueddemann, E. P., 1991, *Silane Coupling Agents*, 2nd ed., New York: Plenum Press.

Pohl, D. W. and Nowotny, L., 1994, *J. Vac. Sci. Technol. B*, 12, 1441.

Sarid, D., 1991, *Scanning Force Microscopy with Applications to Electric, Magnetic and Atomic Forces*, New York: Oxford Univ. Press.

Schürmann, M. and Anselmetti, D., 2012, Bielefeld University, unpublished results.

Snedden, N., 1965, *Inst. J. Eng. Sci.*, 3, 47.

Stöckle, R. M., Suh, Y. D., Deckert, V., and Zenobi, R., 2000, *Chem. Phys. Lett.*, 318, 131.

Sugimoto, Y., Pou, P., Abe, M., Jelinek, P., Pérez, R., Morita, S., and Custance, Ó., 2007, *Nature*, 446, on-line 05530.

Synge, E. H., 1928, *Phil. Mag.*, 6.

Tersoff, J. and Hamann, D. R., 1985, *Phys. Rev. B*, 31, 805.

Tsukruk, V. V. and Blizniuk, V.N., 1998, *Langmuir*, 14, 446.

Turner, D., 1984, U.S. Patent 4,c469, 554.

Ulman, A., (ed.), 1998, *Self-Assembled Monolayers of Thiol*, San Diego, CA: Academic Press.

Vakarelski, I. U. and Higashitani, K., 2006, *Langmuir*, 22, 2931.

van Leemput, L. E. C. and van Kempen, H., 1992, *Rep. Prog. Phys.*, 55, 1165.

Veerman, J. A., Garcia-Barajo, M. F., Kuipers, L., and Van Halst, N. F., 1999, *J. Microsc.,* 194, 477.

Veerman, J. A., Otter, A. M., Kuipers, L., and van Hulst, N. F., 1998, *Appl. Phys. Lett.* 72, 3115.

Watson, G. S., Blach, J. A., Cahill, C., Myhra, S., Nicolau, D. V., Pham, D. K., and Wright, J., 2003, *Coll. Polym. Sci.*, 282, 56.

Wiesendanger, R., Meyer, E., and Morita, S., 2002, *Noncontact Atomic Force Microscopy*, Berlin: Springer.

Wiesendanger, R., (ed.), 1998, *Scanning Probe Microscopy*, Berlin: Springer.

Willemsen, O. H., Snel, M. M. E., Cambi, A., de Groot, B. G., and Figdor, C. G., 2000, *Biophys. J.*, 79, 3267.

Wolkow, R. and Avouris, P., 1988, *Phys. Rev. Letts.*, 60, 1049.

Zhao, X., 2008, *DPhil Project*, UK: University of Oxford.

Zlatanova, J., Lindsay, S. M., and Leuba, S. H., 2000, *Prog. Biophys. Mol. Biol.*, 74, 37.

6 Techniques and Methods for Nanoscale Analysis of Single Particles and Ensembles of Particles

6.1 INTRODUCTION

This chapter focuses on the techniques and methods for characterization of those materials that are assemblies of particles on the nanoscale in three dimensions; more specifically, nanoparticles. Such particles can exist in single, fused, aggregated, or agglomerated forms, with spherical, elongated, or irregular shapes. They may be presented for characterization as powders, aerosols, or after dispersal, in a fluid. They can also be introduced into other materials, such as in a polymer nanocomposite. Obviously, the physical form of a particular nanomaterial has an impact on the choice of the technique(s) and method(s) of its characterization, as well as on handling, safety aspects, and sample preparation. In Chapters 7 and 8, the techniques of choice for characterization of single, or a few, objects that are nanoscale in three dimensions, for example, by techniques such as scanning electron microscopy/transmission electron microscopy (SEM)/(TEM), will be described. An overview of the types, origins, and attributes is given in Figure 6.1. Some types of particles that are of particular current relevance and interest, such as quantum dots, nanotubes, and nanowires, are discussed in other chapters.

The main variables of interest with respect to nanoparticulate characterization are:

Physical properties:
- Size, shape, aspect ratio.
- Agglomeration and aggregation behaviour.
- Surface morphology/topography.
- Structure, including crystallinity and defect structure.
- Solubility.

Chemical properties:
- Bulk elemental/molecular composition.
- Phase identity and purity.
- Surface composition.
- Surface charge.
- Interfacial characteristics.

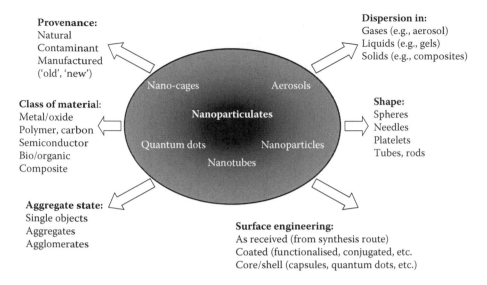

FIGURE 6.1 Characterization parameters for nanoparticulate materials. A distinction is made under the provenance heading, between particles that are manufactured in 'old' (e.g., paints), and 'new' (e.g., quantum dots), industries. (Adapted and redrawn from Schulenburg, 2006.)

Each of these variables may be of significance to properties and performance, and, whilst numerous techniques exist for their measurement, each technique has its own advantages and disadvantages. Measurements that are specific to a particular property may also be necessary; for example, in order to assess the explosive hazard of a nanopowder, it may be necessary to use calorimetric methods to determine ignition temperature and burn rate as a function of particle size.

6.1.1 Particle Size

The size of a nanoparticle is a fundamental parameter to be determined. However, ascertaining the size of a three-dimensional object, whether it is nanosized or not, is a conundrum in that one number will not suffice in general. There is only one 3-D object that can be described by a unique number, and that is a perfect sphere, the number being its radius. The notion of an 'equivalent' sphere has been used to describe nanoparticles, that is, a particle is assumed to approximate to a perfect sphere, and its average dimension is then taken as equivalent to the 'perfect' radius, but this assumption can lead to problems. Equivalent sphere theory assumes that whatever dimension is measured can then be reduced to that of a sphere. Hence, by definition, the theory does not take into account any deviation from a spherical shape, and the 'radius' so derived may therefore be a meaningless parameter. For example, a carbon nanotube (CNT) of radius 1 nm and length 100 nm has an equivalent sphere radius of 4.2 nm! In addition, it must be remembered that the derivation of this parameter does not take into account whether the CNT is of the single-wall or multi-wall variety.

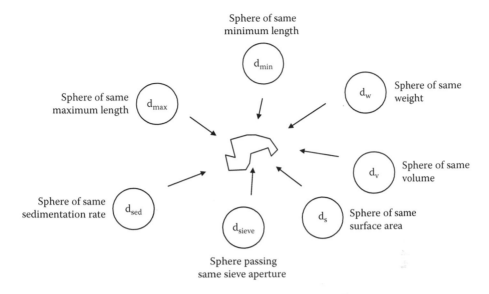

FIGURE 6.2 Equivalent measurements that may be used to describe an irregular particulate. (Adapted and redrawn from Rawle, Basic Principles of Particle Size Analysis, www.malvern.com.)

Furthermore, particulates are rarely monodisperse, and hence it is often necessary to measure the size distribution of particles. This also introduces a number of different ways of reporting similar, but different, data—such as the number, mass, and volume averages. Although these averages are interrelated and can often be converted, the accuracy of the interconversion relies heavily on the statistics of the original measurement. A reliable outcome requires that large numbers of particles must be measured before any interconversion can be attempted. An excellent treatment of the statistics relating to number, mass, and volume counting is presented in Basic Principles of Particle Size Analysis, Malvern Instruments (www.malvern.com). Several equivalent measures are available to describe particulates, as shown in Figure 6.2.

All results, when reported, should outline the measurement method used, to aid comparison, and to allow the reader to draw appropriate conclusions. Even if the particulate is measured in an electron microscope, only those particles that end up on the TEM grid, or SEM stub, will be 'seen'. The average particle size or distribution inferred from TEM/SEM imaging will depend both on the how the particles arrived there, and on the number of particles that have been imaged and measured. Obtaining a statistically sound particle size distribution by this method can be both costly and time-consuming, but it may be the only technique which can give the primary particle size and shape. Bear in mind that even with the TEM/SEM, what is observed is a 2-D projection of the particle (a eucentric tilt stage will improve matters). It is necessary to be aware that information about particle size obtained from different characterization methods may be related to different properties of that particle, and the methods may therefore not all give the same answer. Few techniques (with the exception of TEM) can measure the size of the smallest nanoparticles (ca.

1 nm), so that the latter may well be overlooked by most methods. Similarly, if the nanoparticles are agglomerated, the inferred size distribution may be distorted.

There are many complementary methods for particle sizing of nanoparticles, some of which will be described here.

Particle sizing techniques can be divided roughly into three categories:

(i) **Ensemble methods** collect mixed data from all the differently sized particles in a sample at the same time, and then digest the data to extract a distribution of particle sizes for the entire population. The most common ensemble methods that collect information on nanoparticles dispersed in a liquid are photon correlation spectroscopy (PCS) (also known as 'dynamic light scattering' (DLS)) and backscattering spectroscopy (e.g., see Pecora 1985).

(ii) **Separation methods** all apply an outside separation force to the particles in a distribution in order to separate physically the particles according to size. Since particles of different sizes are actually physically separated, the problems of accurate characterization of individual particles (counting methods) and of calculating a distribution from mixed data (ensemble techniques) are reduced or eliminated. The accuracy of these methods depends on whether the particles react to the separation force as expected, and their effective resolution depends on how completely the particles are separated according to size. Common separation techniques include sieves, gravitational sedimentation, the disc centrifuge, capillary hydrodynamic fractionation, sedimentation field flow fractionation, and others. The disc centrifuge is the method of choice for nanoparticles (for additional details, see www.cpsinstruments.eu).

(iii) **Counting methods** all characterize the sample distribution one particle at a time, basically by accumulating counts of particles with similar sizes. Some common counting methods are: the electrozone counter, the light counter, the time-of-flight counter, and electron or scanning probe microscopies. In each of these methods, particles are classified and placed in 'size bins', one particle at a time. In all cases, it is necessary to ensure that multiple particles are not counted together, and thus cause errors in the reported size distribution (e.g., due to 'co-incident counting'). Their accuracy and resolution depend on how accurately the size of each particle can be characterized during the (usually) very brief time that it is counted. Counting methods will not be described in this section since some, for example, SEM, TEM, and SPM, are described elsewhere in this book (Chapter 2 and Chapter 5, respectively).

6.2 PHOTON-CORRELATION SPECTROSCOPY (PCS) OR DYNAMIC LIGHT SCATTERING (DLS)

For particles or agglomerates of size less than 1000 nm, dynamic light scattering (DLS) is often the method of choice. DLS is based on the dependence of Brownian motion on particle size. Brownian motion is the random movement of particles in a solution due to collisions with the solvent molecules surrounding them. Larger particles move more slowly and over shorter distances, whereas smaller particles move faster and further. Auto-correlation of successive speckle patterns (see below) allows the particle size distribution to be determined within the measurement window; larger particles will move slowly and hence successive speckle patterns will change slowly, whereas small particles will move more quickly and successive patterns will change rapidly. Thus, a correlation function can be constructed which relates to particle size distribution. Temperature stability is an important consideration when performing DLS measurements as liquid viscosity changes with temperature.

6.2.1 Theory

Particle diameters can be calculated from the translational diffusion coefficient using the Stokes–Einstein equation:

$$d(H) = \frac{kT}{3\pi\eta D} \tag{6.1}$$

where $d(H)$ is the hydrodynamic diameter, D the translational diffusion coefficient, k the Boltzmann constant, T the absolute temperature, and η the viscosity.

The diameter so measured is based on the mode of diffusion of a particle within a fluid, and is referred to as the hydrodynamic diameter (see Figure 6.3). It is larger than the true particle diameter as determined by other techniques, for example, by SEM, for which consideration may need to be taken into account since the properties of biological systems are sensitive to the hydrodynamic diameters of particles.

The hydrodynamic diameter of a particle corresponds to the diameter of a sphere having the same translational diffusion as that particle, which depends on the size of the particle core, its surface structure, and the concentrations and types of ions

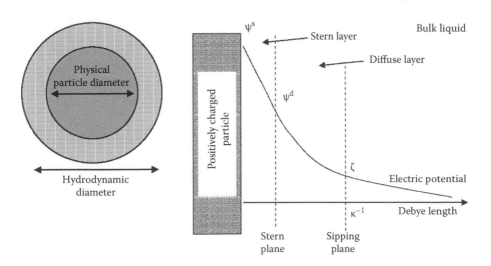

FIGURE 6.3 The hydrodynamic diameter of a particle in solution depends on the Debye length.

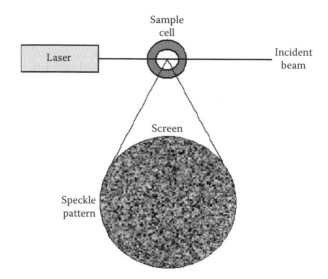

FIGURE 6.4 Schematic representation of a speckle pattern. (From DLS Instrumentation, Dynamic Light Scattering: An Introduction in 30 Minutes, with permission from www.malvern.com.)

in the medium. Factors that affect the measurement of the hydrodynamic diameter include: ionic concentration of the medium, (which changes the electric double layer (defined by the Debye length, κ^{-1})), changes to the surface structure, and extent of non-sphericity.

Then, if the cell containing them is irradiated with suitable laser light and the light scattered from the particles is projected onto a frosted glass screen, a so-called classic speckle pattern would be seen. The arrangement is shown schematically in Figure 6.4. Because the whole system has been assumed to be stationary, the speckle pattern would also be stationary in both speckle size and position. In the pattern, the dark spaces correspond to destructive interference, that is, where the vector additions of the scattered light are out of phase and cancel each other. On the other hand, the bright regions correspond to constructive interference, that is, where the light scattered from the particles arrive in-phase, with consequent phase reinforcement.

When the particles are not stationary, but undergoing Brownian motion, a dynamic speckle pattern would be observed, since the phase addition from the moving particles would then be evolving constantly and forming new patterns. The rate at which these intensity fluctuations occur will depend on the size of the particles. Figure 6.5 illustrates schematically typical intensity fluctuations arising from dispersions of large and small particles, respectively. Small particles cause the intensity to fluctuate more rapidly than large ones.

It is possible to measure directly the spectrum of frequencies contained in the intensity fluctuations arising from the Brownian motion of particles, but it is inefficient to do so. A better way is to use a device called a 'digital auto-correlator'. A correlator is basically a signal comparator. It is designed to measure the degree of similarity between two signals, or one signal with itself at varying time intervals.

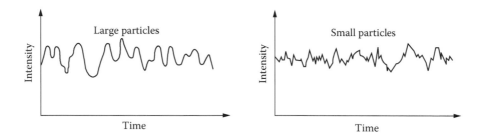

FIGURE 6.5 Typical intensity fluctuations for large and small particles, respectively. (From DLS Instrumentation, Dynamic Light Scattering: An Introduction in 30 Minutes, with permission from www.malvern.com.)

If the intensity of a signal is compared with itself at a particular point in time, and then again, at a much later time, it is obvious that for a randomly fluctuating signal the intensities will not be related in any way, that is, there will be no correlation between the two signals. Knowledge of the initial signal intensity does not allow the signal intensity at time t = ∞ to be predicted, which will be true of any random process such as diffusion.

However, if the intensity of a signal at time t is compared to the intensity of the same signal at an incremental time later (t + δt), when δt is significantly shorter than the time-scale of evolution of the random dynamics of the system under study, there will be a strong relationship or correlation between the intensities of two signals.

If the signal at t, derived from a random process such as Brownian motion, is compared to the signal at t + 2δt, there will still be a reasonable correlation between the two signals, but it will not be as good as that at t and t + δt. The strength of the correlation reduces with time. The period of time δt is usually very small, perhaps ns or μs, and is called the 'sample time of the correlator,' while t = ∞ may be of the order of 1–10 ms.

If the signal intensity at t is compared with itself, then there is perfect correlation as the signals are identical. Perfect correlation is indicated by unity (1.00) and no correlation is indicated by zero (0.00). If the signals at t + 2δt, t + 3δt, t + 4δt, and so forth are compared with the signal at t, the correlation of a signal arriving from a random source will decrease with time until at some time, effectively t = ∞, there will be no correlation. A plot of correlation coefficient versus time is called a '*correlellogram*', as in Figure 6.6. If the particles are large, the signal will change slowly and the correlation will persist for a long time (b), but if they are small and moving rapidly, then the correlation will reduce more quickly (a).

Simple examination of a correlellogram can provide a great deal of information about the sample. The time at which the correlation starts to decay significantly is an indication of the mean size of the particles making up the sample, while greater steepness of the slope of the decay line correlates with greater monodispersivity of the particle distribution. Conversely, the more extended the decay, the greater the polydispersity of the particle distribution.

Particles in a dispersion are in constant random Brownian motion, causing the intensity of scattered light to fluctuate as a function of time. The correlator used in

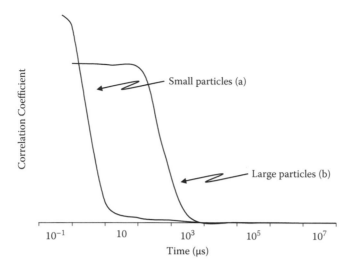

FIGURE 6.6 Typical correlellogram from (b) a sample containing small particles where the correlation signal takes a long time to decay, and (a) a sample containing large particles, where the correlation signal decays more rapidly. (Adapted from www.malvern.com, with permission.)

a PCS instrument constructs the correlation function $G(\tau)$ of the scattered intensity, according to

$$G(\tau) = \langle I(t) \cdot I(t + \tau) \rangle \qquad (6.2)$$

where τ is the time difference (the sample time treated as a continuous variable) of the correlator.

For a large number of monodisperse particles in Brownian motion, the correlation function is an exponentially decaying function of the correlator time delay τ:

$$G(\tau) = A[1 + B \exp^{-\Gamma r}] \qquad (6.3)$$

where A is the baseline [$G(\tau)$ for $\tau \to \infty$] of the correlation function, and B is the intercept on the ordinate axis at $\tau = 0$ of the correlation function. The decay parameter is defined by

$$\Gamma = Dq^2 \qquad (6.4)$$

where D is the translational diffusion coefficient, and q is given by

$$q = \left(\frac{4\pi n}{\lambda_0} \right) \sin\left(\frac{\theta}{2} \right) \qquad (6.5)$$

where n is the refractive index of the dispersant, λ_0 is the wavelength of the laser, and θ is the scattering angle.

For polydisperse particle distributions, Equation (6.3) can be written as:

$$G(\tau) = A[1 + Bg_1(\tau)^2] \tag{6.6}$$

where $g_1(\tau)$ is the sum of all the exponential decay terms contained in the correlation function.

Size can be obtained from the correlation function by using various algorithms. There are two approaches that can be taken, viz.,

(i) Fit a single exponential to the correlation function to obtain the mean size (z-average diameter) and an estimate of the width of the distribution (polydispersity index) (this is called the 'Cumulants analysis' and is defined in ISO13321, Part 8).

(ii) Fit multiple exponential terms to the correlation function to obtain the distribution of particle sizes (using software such as Non-Negative Least Squares (NNLS) or (CONTIN)).

The size distribution so obtained is a plot of the relative intensity of the light scattered by particles in various size classes, and is therefore known as an intensity/size distribution.

If the distribution by intensity is a single fairly smooth peak, then there is little point in carrying out the conversion to a volume distribution using the procedure described above (www.malvern.com). If the optical parameters are correct, that will merely provide a peak of slightly different shape. However, if the plot shows a substantial tail, or more than one peak, then, using appropriate refractive indices for the sample, the intensity distribution can be converted to a volume distribution, which will give a more realistic view of the importance of the tail or of the presence of a second peak. In general terms, it is found that the particle size, d, derived from different methods of calculation will vary as follows:

$$d_{intensity} > d_{volume} > d_{number} \tag{6.7}$$

It is important to note that large particles make a disproportionate contribution to volume, in comparison with small particles. Also, Rayleigh scattering from particles has a $1/d^6$ dependence so small particles are swamped by larger particles in terms of scattered intensity.

A simple way of illustrating the difference between intensity, volume, and number distributions is to consider two populations of spherical particles of diameter 5 and 50 nm, respectively, present in equal numbers (Figure 6.7). If a number distribution of the two particle populations is plotted, then the plot should consist of two peaks (positioned at 5 and 50 nm) in a ratio of unity. If this number distribution were to be converted into a volume distribution, then the ratio of the two peaks would change to 1:1,000 (because the volume of a sphere is proportional to $(d/2)^3$). If this were converted further into an intensity distribution, a 1:1,000,000 ratio between the two peaks would be found (because the intensity of scattering is proportional to d^6

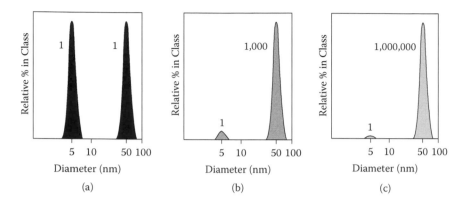

FIGURE 6.7 (a) Number, (b) volume, and (c) intensity, distributions of a bimodal mixture of 5 and 50 nm particles present in equal numbers. (Adapted from www.malvern.com, with permission.)

(from Rayleigh's approximation)). The significance is that in DLS, the distribution obtained from a measurement is based on intensity.

6.2.3 MIE THEORY OF SCATTERING

In contrast to traditional diffraction theory, the scattered intensity becomes a complex function of particle size with a series of maxima and minima, when that size approaches the wavelength of light. The application of Mie (or Lorenz–Mie) scattering theory (Mie 1908; van de Hulst 1957) embraces all possible ratios of diameter, d, to wavelength, λ, and solves the equations of classical electromagnetism for the interaction of light with matter. As a special case, it has been shown (Doremus 2002) that, in the limit of very small spherical conductive particles ($d \ll \lambda$) embedded in a medium of refractive index n, coupling to the free electron plasma gives rise to optical absorption given by the expression

$$\alpha = \frac{18\pi Q n^3 \varepsilon_2 / \lambda}{(\varepsilon_1 + 2n^2)^2 + \varepsilon_2^2} \tag{6.8}$$

where Q is the volume fraction of particles, and ε_1 and ε_2 are the real and imaginary parts of the complex dielectric constant. As a result, there will be a $1/\lambda^4$ dependence on absorption in the limit of $d \ll \lambda$.

Rayleigh theory predicts a $1/\lambda^6$ dependence for the scattered intensity and a weak angular dependence for the direction of incident light with respect to the direction of detection. A similar result is obtained for $d \ll \lambda$ as the limiting case of the Mie theory. When the diameter of the scattering particle is comparable to, or greater than, the wavelength of the incident radiation, then Mie theory predicts a much stronger angular dependence with forward scattering being favoured, as well as pronounced oscillations in the scattering cross-section as a function of scattering angle, and in the scattered intensity as a function of particle size.

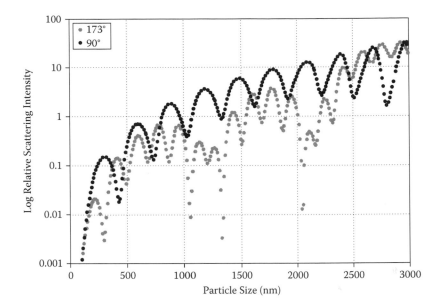

FIGURE 6.8 Theoretical plot of the log of the relative intensity of scattering versus particle size at scattering angles of 173° (upper trace) and 90° (lower trace), assuming a laser beam at a wavelength of 633 nm, real refractive index of 1.59 and an imaginary refractive index of 0.001. (Adapted from www.malvern.com, with permission.)

In most cases, data reduction of results obtained by optical scattering based on Mie theory provides accurate characterization over a range from ca. 10 nm upwards. It does, however, assume that the refractive indices of the material and the medium are accurately known and that the absorption part of the refractive index is also known (an informed estimate may suffice). For the majority of cases, these values are either known or can be measured. Figure 6.8 shows the theoretical plot of the log of the relative scattering intensity versus particle size at scattering angles of 173° and 90°.

6.2.4 DLS INSTRUMENTATION

A typical DLS system consists of six main modules as shown in Figure 6.9. Firstly, a laser (1) provides a light source to illuminate the sample contained in a cell (2). For dilute concentrations, most of the laser beam passes through the sample, but some is scattered by the particles within the sample into all angles. A detector is used to measure the scattered light. In one of the popular instruments, the detector position can be at either 173° or 90° (3), depending on the particular mode to be used for data reduction (www.malvern.com).

The intensity of the scattered light must be within a specific range for optimal sensing by the detector. If that intensity is too high, the detector will become saturated, in which case an attenuator (4) can be used to reduce the intensity of the laser source and hence reduce the intensity of scattered light. For samples that do

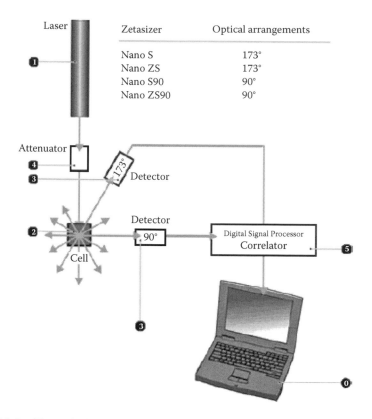

FIGURE 6.9 The optical configuration for DLS. (Zetasizer block diagram, 2007, with permission from www.malvern.com.)

not scatter much light, such as those with particles that are either very small or in low concentration, the amount of scattered light must be increased. In that case, the attenuator will allow higher laser light intensity through to the sample. For some models, the appropriate attenuation is determined automatically by software and covers a transmission range of 100% to 0.0003%.

The scattering intensity signal from the detector is passed to a digital processing board where a correlator (5) compares the scattered intensity at successive time intervals in order to derive the rate at which the intensity is varying. The information from it is then passed to a computer (6), where software analyzes the data and derives size information.

The advantages and disadvantages of DLS can be summarised as follows:

Advantages:
- A minimal amount of information about the sample is needed to run an analysis. Even mixtures of different materials can be accurately measured; only the viscosity of the medium must be known accurately.
- Very small minimum measurable particle size, ca. less than 10 nm.
- Only a small sample is needed, less than 10 µl or less than 0.001 mg.

- The analysis is fast and simple.
- Testing is non-destructive, so samples can be recovered if needed.
- The instrumentation can be coupled to flow cells for automated measurements of the zeta potential.

Disadvantages:

- Low resolution; particles must differ in size by 50% or more for DLS to discriminate between two monodisperse populations. The method does not really provide much 'size distribution' data, only a mean size and estimate of standard deviation.
- A small quantity of a small size particle can easily be 'lost' in a much larger quantity of a large size particle.

6.3 DIFFERENTIAL CENTRIFUGAL SEDIMENTATION (DCS)

There are a number of separation methods but here the emphasis will be on differential centrifugal sedimentation (DCS), a conceptually simple technique, which has been 'reborn' in recent years. Previous limitations and difficulties with the technique of sedimentation have been overcome with recent advances in technology, and DCS is now a versatile tool for the measurement of nanoparticle size distributions down to 3 nm. With its unique ability to resolve multi-modal particle distributions of closely spaced sizes, to within 2%, and to distinguish extremely small shifts and changes in particle size, DCS is once more becoming a valuable particle characterization tool. The practical range of the technique is from about 3 nm up to about 80 μm (the exact range will be dependent on particle density). The technique is relatively user-friendly, is highly accurate and reproducible, can measure up to 40 samples in the same 'run', does 'speed ramping' for measurement of broad distributions in a single sample, and can even measure 'buoyant' or 'neutral density' particles, (i.e., particles having a lower density than that of the medium in which they are dispersed). Due to the high resolution achievable, DCS is ideal for resolving aggregates and agglomerates, and for observing very small relative shifts in peaks and tails of particle size distributions. It may also be used to measure absolute particle size; however, the density of the particulate material must be known. It can be used for quantitative measurements if the refractive index of the particulate is known. Number or weight distributions can also be easily calculated and displayed. Figure 6.10 shows a typical multi-particle measurement.

6.3.1 THE BASICS OF DIFFERENTIAL SEDIMENTATION

Sedimentation of particles in a fluid has long been used to characterize particle size distribution. It is based on the use of Stokes' Law for the determination of an unknown distribution of spherical particle sizes, by measuring the time required for the particles to settle a known distance in a fluid of known viscosity and density. Sedimentation can be either gravitational (under a 1 G force), or centrifugal (under a many-G force). For a centrifuge running at constant speed and temperature, all the parameters in the expression for Stokes' Law, except time, are constant during an analysis. Their values will be either well known or can be accurately measured.

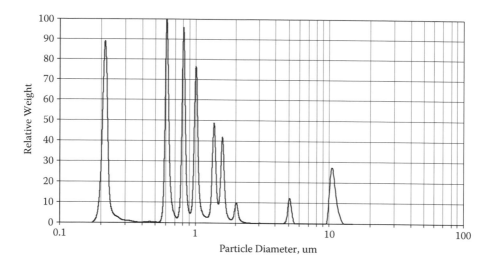

FIGURE 6.10 A trace showing a typical example of a multi-modal particle distribution measured by DCS. Nine polystyrene standards, all of different sizes, were mixed and injected into the instrument. Each particle size can be seen to be resolved clearly. (With permission from Hiran Vegard, www.cpsinstruments.eu/pdf/General%20Brochure.pdf.)

Within a broad range of analysis conditions, a modified form of Stokes' Law can be used to measure accurately the diameter of spherical particles, based on their arrival time at the detector. Hence, by introducing a known, traceable standard, the time scale can be calibrated against particle size, via the expression

$$V = \frac{2D^2(\rho_P - \rho_F)G}{9\eta} \qquad (6.9)$$

where D is the particle diameter (m), ρ_P the particle density (kg/m³), ρ_F the fluid density (kg/m³), G the gravitational acceleration (m/sec²), and η the fluid viscosity (Pa-s).

6.3.2 DCS Instrument Design

The most common design for DCS instruments is that of a hollow, optically clear, disc mounted vertically and driven by a variable speed motor. A typical front view and disc cross-section are shown in Figure 6.11. The outer reinforcing ring is made of Kevlar-reinforced aluminum to withstand the high-G-force when the disc is spinning at high speed (up to 24,000 rpm). The disc has a central closure cap, which can be opened when the contents need to be emptied. When the disc is rotating at speed, any fluid in it is forced centrifugally to the outer edge. The amount of fluid through which sedimentation is to occur corresponds to a sedimentation depth of 1–2 cm.

For the above-mentioned sedimentation fluid, rather than using pure water, a so-called gradient fluid (see Figure 6.11) is used, in which a gradated density profile is

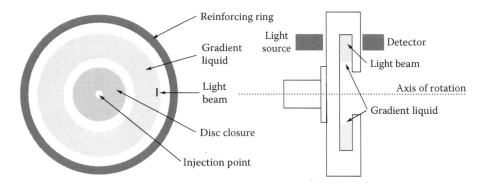

FIGURE 6.11   A typical disc from a DSC instrument. Front view and cross-section, illustrating how a liquid with a density gradient would fill the hollow section during rotation. The optically clear disc is mounted vertically as shown on the right, and driven by a variable speed motor. (With permission from Hiran Vegard, www.cpsinstruments.eu.)

achieved by the use of, for example, solutions of sucrose in water, typically 8% and 24%. Despite difficulties in the concept of density gradients in miscible liquids, the gradient can, in fact, be set up in the following way. First, the empty disc is set in motion to the chosen maximum rotational speed; then, the sucrose/water solutions are injected along the axis of the spinning disc, with the solution(s) of lower concentration being injected before those of higher concentrations. As long as the disc remains spinning at high speed, the solutions will not mix appreciably, and a density gradient is maintained that increases more or less continuously from inside to the outer edge. Of course, when the disc stops, the gradient is destroyed. Injection can be carried out either by an automatic 'gradient builder', or manually.

If such a density gradient is not used, then a sample of dispersed particles in suspension would simply pass through the sedimentation fluid without appreciable particle separation. This is commonly called '*streaming*' or '*sedimentation instability*', and results in a broad particle-size distribution peak rather than in separate peaks for each band of particle size, resulting in the loss of all information about the particle-size distribution. Streaming was a common problem in older instrumentation which used integral rather than differential sedimentation (i.e., no density gradient).

The disc can be of virtually any size, but manufacturers have settled on a diameter of about 125 to 150 mm. A light source, usually monochromatic light of relatively short wavelength (400–500 nm), and a detector, are placed on opposite sides of the moving disc so that there is continuous irradiation through the fluid near the edge of the disc. Although some instruments use a longer wavelength (650 nm), or X-rays, it has been accepted that near-ultraviolet (UV) wavelength light gives better detector sensitivity when particles smaller than 100 nm are to be measured. At first, the detector indicates maximum intensity, but as particles undergo sedimentation through the gradient fluid and reach the measurement sector, the signal reduces. The reduction in intensity is related to the concentration of particles in the detector beam (the Mie theory of light scattering can be applied here). When all the particles have passed the detector, the signal returns to its original level. The raw data (time versus intensity) can be used to draw a plot of particle concentration against calculated particle diameter, once a National Institute of Standards and Technology (NIST) traceable standard has been used to calibrate sedimentation time (plotted on the x-axis) versus particle size.

Samples are prepared for analysis by dilution in a fluid (normally water) of density slightly lower than that of the least dense fluid in the disc. When a sample (of volume normally around 100 µl) is injected at the centre of the moving disc, it first strikes the back inside face of the disc, and then forms a thin film, which spreads as it accelerates radially toward the surface of the gradient fluid. When the dispersed sample solution reaches the gradient fluid surface, it quickly spreads over that surface because of its lower density. In fact, the sample solution 'floats' on the gradient fluid, and the individual particles begin immediately to be driven centrifugally through the gradient to the outer edge, passing through the detector beam, on their way. The injection of a sample is rapid (typically <50 ms), so the starting time for an analysis is well defined, and the precision of sedimentation time is correspondingly good. When an analysis is complete, all the particles will have passed through the detector beam and end up accumulated at the edge of the disc, where they will stay for as long as the disc is in motion. Since the dispersed particles have at that stage all passed through the gradient fluid, the instrument is then ready for the next sample. There is no need to empty and clean the disc; therefore, many samples can be run in sequence without stopping it.

A calibration standard, which has a known particle size and particle density and is NIST traceable, is run before the sample, or optionally in between samples, so that the instrument can be calibrated, enabling sample analysis time to be converted to particle size. Figure 6.12 shows a centrifuge disc in an instrument during operation.

Discs have been developed that enable other types of analysis to be performed that were previously very difficult. For example, when particles have a lower density

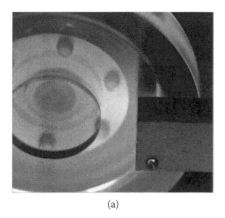

(a)   (b)

FIGURE 6.12 The images show (a) a centrifuge disc inside an instrument and the light source-detector towards the outside of the disc, and (b) the same disc in rotation during an analysis, where the separated bands of differently sized particles can be seen clearly as they approach the detector towards the outside of the disc. (With permission from Hiran Vegard, www.cpsinstruments.eu.)

than the medium in which they are dispersed, they have a tendency to float rather than settle. Special low-density discs, combined with reversal of the detector position, can now allow these types of samples to be measured.

6.4 ZETA POTENTIAL

Zeta potential refers to the electrostatic potential generated by the accumulation of anions at the surface of a colloidal particle in a fluid medium; the potential is across an electrical double-layer consisting of the Stern layer and the diffuse layer (as shown in Figure 6.3). The zeta potential of a particle can be calculated, if the electrophoretic mobility of the sample is known, by Henry's equation, which is:

$$U_e = \frac{2\varepsilon \zeta f(ka)}{3\eta} \tag{6.10}$$

where U_e is the electrophoretic mobility, ε the dielectric constant of the sample, ζ the zeta potential, $f(ka)$ Henry's function (most often the Hückel and Smoluchowski approximations of 1 and 1.5, respectively, are used), and η the viscosity of the solvent medium.

The zeta potential is easily measured using a laser doppler velocimeter (LDV). An electrical field of known strength is applied across a colloidal sample, through which a laser beam aligned with the direction of ionic drift is then passed. The electrophoretic mobility of the colloid dictates the velocity of the charged particles, which in turn induces a frequency shift in the incident laser beam. Using the dielectric constant of the sample, the viscosity of the solvent, the measured electrophoretic mobility, and either the Hückel or the Smoluchowski approximation for Henry's function, the zeta potential of the particles within the colloid can be calculated. It is

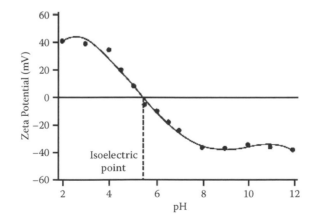

FIGURE 6.13 A plot showing an isoelectric point at pH5.5. It can be predicted that between pH values of <4 and >7.5 stability problems will arise since the zeta potential values are numerically less than 30 mV. (From Zeta Potential: An Introduction in 30 minutes, Downloadable application note, with permission from www. malvern.com.)

obvious that PCS/DLS can also be used to measure the zeta potential from changes in the hydrodynamic particle size as a function of, for example, ionic strength, or more commonly pH.

The importance of the zeta potential of a colloid is that of a relative measure of the stability of the system. The DLVO theory (Derjaguin and Landau 1941; Verwey and Overbeck 1948) for colloidal interactions dictates that a colloidal system will remain stable, if, and only if, the Coulombic repulsion arising from the net charge on the surface of the particles in a colloid is greater than the van der Waals force between those same particles. When the reverse is true, the colloidal particles will cluster together and form flocculates and aggregates (depending on the strength of the van der Waals attraction and the presence/absence of steric effects). The higher the absolute zeta potential, the stronger the Coulombic repulsion between the particles within a colloid, and, therefore, the less the impact of the van der Waals force on the colloid. The line dividing stable and unstable suspensions is usually taken to be around ± 30 mV (numerically greater than 30 mV and the suspension would normally be considered stable).

Factors that affect stability include pH, conductivity, and concentration of ions in solution. For example, the addition of more alkali will induce a more negative charge, and vice versa with addition of acid. Figure 6.13 shows a typical plot of zeta potential against pH.

The thickness of the double layer depends on the concentration of ions in solution, and can be calculated for any particular ionic strength. High ionic strength will compress the double layer as will the valency (e.g., Fe^{3+} versus Na^+). Specific adsorption of ions onto the surface can have a dramatic effect on the isoelectric point, which may lead to complete reversal of the surface charge. Colloidal concentration can also affect the zeta potential. The effects described above can be used to great advantage during formulations to resist, for example, flocculation.

6.5   DIFFERENTIAL MOBILITY SPECTROMETRY (DMS)

This technique is appropriate to particles, in a range of sizes down to molecular, in gaseous suspension (e.g., an aerosol). The particles are ionized in a preparation chamber, usually by a radioactive source, and then injected into a drift chamber where they come under the influence of a direct current (DC) bias and a super-imposed radio frequency (RF) field. Different size fractions of particles are separated due to non-linear differences in mobility at high and low field strength for ions of different size and charge in the carrier gas, and are then sorted according to the time delay of arrival at the collector plates, where the point of injection constitutes time t = 0. The schematic of a typical instrumental configuration is shown in Figure 6.14.

6.6   SURFACE AREA DETERMINATION

The surface area of a solid material is the total area of the material that is in contact with the external environment, and is generally expressed as m^2/gm of dry sample. In many materials, this area is not merely that of the external, visible, surface, but

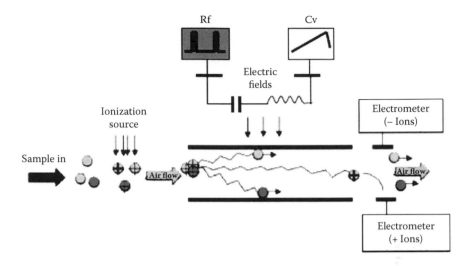

FIGURE 6.14 (**See color insert.**) Schematic of a differential mobility spectrometer (DMS). (Adapted from Davis, 2009, *Intl. Gases Instrum.*)

includes that of the internal surfaces as well. Materials in the forms of both solid and powder, either of natural origin (e.g., stones, soils, minerals, etc.) or of industrial origin (e.g., catalysts, pharmaceuticals, metal oxides, ceramics, carbons, zeolites, etc.) can contain internal void volumes. These are distributed within the solid in the forms of pores, cavities, and cracks, and the total sum of the volumes is called the 'porosity'. Porosity is a strong determinant of certain important physical properties of materials, such as durability, mechanical strength, permeability, adsorption behaviour, and so forth, and knowledge about pore structure, particularly the internal surface area, is important in characterizing materials and predicting their behaviour. When significant porosity is present, the ratio of the external surface area to that of the total surface area may well be quite small. Measurement of surface area can be made from an adsorption isotherm (see below).

There are two principal types of pore: closed and open. Closed pores are isolated completely from the external surface, and direct access by liquids and gases is not possible. However, closed pores influence mechanical and thermal properties as well as density. Open pores are by definition connected to the external surface, and are therefore accessible to fluids and gases, depending on the pore nature and size and on the particular fluid or gas. They can be subdivided further into terminated and interconnected pores.

The characterization of porosity in solids consists of the determination of the average pore size. Pore dimensions cover a very wide range, and can be classified into three main groups:

- Micropores: diameters less than 2 nm.
- Mesopores: diameters between 2 and 50 nm.
- Macropores: diameters greater than 50 nm.

6.6.1 ADSORPTION AND DESORPTION AT SURFACES

Adsorption occurs when molecules from the gas phase impinge on a surface and remain attached to it. It can take two forms, depending on the strength of the bond attaching the molecules to the surface. If there is chemical reaction between a molecule and the surface, involving the transfer of electrons and the resultant formation of a chemical bond, then the process is called *'chemisorption'*. In the other form, no chemical reaction is involved, but molecules approaching a surface experience a weak attractive force due to the intrinsic surface energy. Such a force is called the *'van der Waals'* force, and the process is known as *'physisorption'*. Chemisorbed layers are difficult to remove, because of the strength of the bond, whereas physisorbed layers can be desorbed easily by an increase in temperature. Physisorption is therefore a reversible phenomenon. Because of this temperature dependence, experiments in physisorption are normally carried out at a low temperature, most conveniently that of liquid nitrogen.

In chemisorption, the surface reaction is site-selective, and the course of the reaction can be complicated, with the possible formation of islands of reaction products, in which there are several successive reacted layers. On the other hand, in physisorption, the coverage of the surface occurs in a consecutive layer-by-layer fashion, usually but not always with the first adsorbed layer being complete before the second starts on top of it, and is not site-selective. Thus, to a good approximation, the amount of gas adsorbed in the first layer is proportional to the surface area of the solid that is available to the gas. That area will consist of both external and internal (e.g., pore) surface areas, assuming that the pore sizes allow ingress of the gas molecules. This proportionality forms the basis of a widely used and well-established method for the measurement of surface area, called the *'BET'* (Brunauer, Emmett, and Teller 1938) method.

The BET method is based on the understanding of the Langmuir and BET isotherms. The derivation carried out below with minor changes in notation, closely follows a version in the literature (de Boer 1953), which was based on the original work by the BET team (Brunauer, Emmett, and Teller 1938).

Molecules adsorb, or desorb, to/from a surface. Adsorption can be described in terms of:

- The fraction of the bare surface area that is covered by adsorbate molecules in a layer of monomolecular thickness, θ_1.
- The fraction of the surface area of the first adsorbed layer that is covered by a second layer of molecules, θ_2.
- The fraction of the surface area of the second adsorbed layer that is covered by a third layer of molecules, θ_3.
- Extension of the sequence to additional layers, that is, the fractional surface coverage for the ith layer is θ_i.

The total number of molecules adsorbed per unit area, A, is therefore

$$A = a_0\theta_1 + 2a_0\theta_2 + \dot{g}\dot{g}\dot{g} + ia_0\theta_i + \dot{g}\dot{g}\dot{g} \qquad (6.11)$$

where a_0 is the number of molecules that would cover a unit area of the surface with a continuous monomolecular layer. Equation (6.11) can be written as

$$A = a_0 \sum_{i=1}^{i=\infty} i\theta_i \qquad (6.12)$$

The system will reach a steady state, in which all the fractional coverages, including that on the bare surface, θ_0, become constant. Thus,

$$\theta_0 = 1 - (\theta_1 + \theta_2 + \dot{g}\dot{g}\dot{g} + \theta_i + \dot{g}\dot{g}\dot{g}) = 1 - \sum_{i=1}^{i=\infty} \theta_i \qquad (6.13)$$

In this state, the rate of adsorption (+) per unit area on the bare surface, k_{+0}, must be balanced by the rate of desorption (−) per unit area from the surface that has monomolecular coverage, k_{-1}. Accordingly,

$$k_{+0}\theta_0 = k_{-1}\theta_1 \qquad (6.14)$$

The steady state condition for the first adsorbed layer can be described similarly in terms of a balance between adsorption on the bare surface and on the first layer, and desorption from the first and second layers, that is,

$$k_{+0}\theta_0 + k_{-2}\theta_2 = k_{-1}\theta_1 + k_{+1}\theta_1 \qquad (6.15)$$

Applying the condition in Equation (6.14), gives

$$k_{-2}\theta_2 = k_{+1}\theta_1 \qquad (6.16)$$

The argument can be extended readily to subsequent layers, that is,

$$k_{-i}\theta_i = k_{+(i-1)}\theta_{i-1} \qquad (6.17)$$

At this point, it is useful to introduce the simplifying, and reasonable, assumption that the various rate constants for adsorption are identical (except for adsorption on the original surface, which is different from all subsequent layers), and likewise that all rate constants for desorption are equal. Then a parameter x, equal to the ratio of the rate constants, that describes adsorption and desorption from/to all layers beyond the original bare surface, can be defined as

$$k_{+(i-1)} / k_{-i} = x \qquad (6.18)$$

Using the above assumption, a set of simplified expressions can be written as

$$\theta_2 = x\theta_1$$

$$\theta_3 = x\theta_2 = x^2\theta_1 \qquad\qquad (6.19)$$

$$\theta_i = x^{i-1}\theta_1$$

This simplification gives the following expressions for equation (6.12):

$$A = a_0 \sum_{i=1}^{i=\infty} i\theta_i = a_0 \sum_{i=1}^{i=\infty} ix^{i-1}\theta_1$$

$$(6.20)$$

$$= a_0 \frac{k_{+0}}{k_{+1}} \theta_0 \sum_{i=1}^{i=\infty} ix^i$$

The latter expression has been derived by substitution from equation (6.14), (in the form $\theta_1 = (k_0/k_{-1})\theta_0$), and then by multiplying, and dividing, by $x = k_{+1}/k_{-1}$, in order to arrive at a sum that can readily be evaluated, and in order to show explicitly that the differences in rate constants for the first and subsequent layers have been accounted for. Note that k_{+0} is the rate constant for adsorption on the original surface, and k_{+1} is the rate constant for adsorption on the subsequent layers. Also, from Equation (6.13)

$$\theta_0 = 1 - \sum_{i=1}^{i=\infty} \theta_i = 1 - \frac{k_{+0}}{k_{+1}} \theta_0 \sum_{i=1}^{i=\infty} x^i$$

and so

$$\theta_0 = \frac{1}{1 + \dfrac{k_{+0}}{k_{+1}} \displaystyle\sum_{i=1}^{i=\infty} x^i} \qquad\qquad (6.21)$$

In the interest of making the expressions more compact, it is useful to define $C = k_{+0}/k_{+1}$. The summations in Equation (6.21) can be evaluated readily as sums of geometrical progressions,

$$\sum_{i=1}^{i=\infty} x^i = \frac{x}{1-x}$$

$$(6.22)$$

$$\sum_{i=1}^{i=\infty} ix^i = \frac{x}{(1-x)^2}$$

Substitutions of the respective sum expressions, Equation (6.22), into Equations (6.20) and (6.21), followed by combining the resulting equations, with C as a new parameter, give a more compact expression for A. Rearrangement gives the following form

$$A = \frac{Ca_0 x}{(1-x)(1-x+Cx)} \tag{6.23}$$

The underlying theory of the BET treatment of surface adsorption is represented by Equation (6.23). Various forms of the expression can now be used to describe the results of its technical implementation (the notation can vary!).

It can be shown from the kinetic theory of gases that the number of molecules, n, in a gas striking a unit area of surface per unit time is given by

$$n = \frac{Np}{\sqrt{2\pi mRT}} = \beta p \tag{6.24}$$

where N is Avogadro's number, m is the molecular mass, p is the pressure, R is the gas constant and T is the temperature; the constants are lumped into a parameter β, making it explicit that n is directly proportional to the pressure. At constant temperature, the rate of adsorption, and thus the total molecular coverage, must be proportional to the rate of arrival of molecules, which is proportional to pressure. Accordingly, the parameter x must be proportional to pressure. In accord with common practice, it is convenient to define yet another parameter q, so that $x = p/q$. This leads to another version of the BET expression,

$$A = \frac{Ca_0 p}{(q-p)(1+(C-1)p/q)} \tag{6.25}$$

The merit of this version is that the condition $p = q$ can be thought of as a saturation pressure, p_o, where all layers will be filled to capacity.

An alternative and equivalent expression, in the notation of Brunauer (1943), describes the isotherm in terms of the volume of adsorbed gas,

$$v = \frac{Cv_m p}{(q-p)(1+(C-1)p/q)} \tag{6.26}$$

where v_m is the volume of gas required for adsorption of a complete monomolecular layer. A more practical version of the expression for the BET isotherm requires that the parameter q be associated with the saturation pressure, p_0, as suggested above. Making that substitution in Equation (6.26), and doing some further rearranging, leads to the following form:

$$\frac{p}{v(p_0 - p)} = \frac{1}{v_m C} + \frac{C-1}{v_m C}\frac{p}{p_0} \tag{6.27}$$

This version is simply the equation for a straight line if

$$\frac{p}{v(p_0 - p)}$$

is plotted against

$$\frac{p}{p_0}$$

The BET isotherm reduces to the Langmuir isotherm when the pressure is low, that is, when $x \ll 1$. Equation (6.25) rewritten with $A/a_0 = \theta$ then reduces to

$$\theta = \frac{Cx}{1 + Cx} \tag{6.28}$$

which describes the Langmuir situation where only the initial monomolecular layer is considered, in accord with the following conditions.

- Adsorption is restricted to one monolayer.
- All available sites are equivalent.
- There is no interaction between the adsorbed species.

The two isotherms, Langmuir (lower curve) and BET (upper curve) are compared in Figure 6.15.

6.6.2 Surface Area Measurement by the BET Method

The BET method is an extension of the Langmuir theory (Langmuir 1919), which deals with monolayer molecular adsorption leading to multilayer adsorption based on the following hypotheses: (a) physical adsorption of gas molecules on a solid occurs indefinitely in a layer-by–layer manner; (b) there is no interaction between each adsorption layer; and, (c) the Langmuir theory can be applied to each layer. The BET Equation (6.27) can be written as:

$$\frac{1}{v\left(\frac{p_0}{p} - 1\right)} = \frac{1}{v_m C} + \frac{p(C-1)}{p_0 v_m C} \tag{6.29}$$

where p and p_0 are the equilibrium and the saturation pressures, respectively, of an adsorbate at the temperature of adsorption.

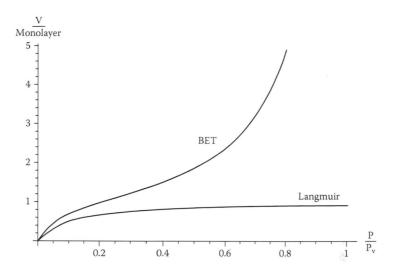

FIGURE 6.15 A schematic comparison of the Langmuir and BET isotherms.

Equation (6.29) represents an adsorption isotherm and can be plotted as a straight line with $1/[v[1 - (p_0/p)]]$ on the y-axis and p/p_0 on the x-axis. This plot is called a 'BET plot,' and an example is shown in Figure 6.16. The linear relationship of this equation is maintained only in the range $0.05 < p/p_0 < 0.35$. The value of the slope A and the y-intercept I of the line are used to calculate the monolayer adsorbed gas quantity v and the BET constant C. The following expressions can be used:

$$v = \frac{1}{A+I} \quad \text{and} \quad C = 1 + \frac{A}{I} \tag{6.30}$$

A total surface area S_{total} and a specific surface area S can be evaluated from the following equations:

$$S_{total} = \frac{vNs}{v_m} \quad \text{and} \quad s = \frac{S_{total}}{MW} \tag{6.31}$$

where N is Avogadro's number, s is the adsorption cross-section, v is the molar volume of adsorbent gas, and MW is the molar weight of adsorbed species.

The most common analysis based on the adsorption isotherm uses the multi-point BET method to measure total surface area. A typical isotherm for a porous material, (a zeolite catalyst), is shown in Figure 6.17. The large uptake of N_2 at low p/p_0 indicates filling of the smallest pores (of size < 2 nm) in the catalyst. The linear portion of the curve represents multilayer adsorption of N_2 on the surface of the catalyst, and the concave upward portion of the curve corresponds to the filling of meso- (of size 2–50 nm) and macropores (of size >50 nm). An entire isotherm is needed to calculate the pore size distribution of the catalyst. However, for a surface area evaluation, data in the relative pressure range 0.05–0.30 are generally used.

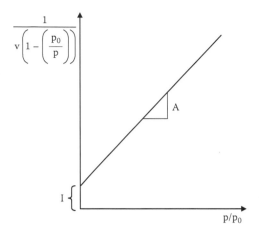

FIGURE 6.16 Schematic of BET plot. (Adapted from http://en.wikipedia.org/wiki/File:BET-1.jpg.)

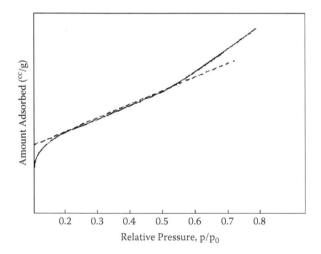

FIGURE 6.17 A typical nitrogen adsorption isotherm, obtained for a zeolite. (From http://www.refiningonline.com/engelharddkb/crep/TCR2_10.htm.)

Modern instruments usually measure volume and pressure and relate them to a modified BET equation, in which theory predicts a linear relationship between slope A and intercept I when the pressure of the adsorbate gas p, saturation vapour pressure p_o, and volume of gas adsorbed v is plotted against the relative pressure, as in

$$\frac{p}{v(p_o - p)} = A\,\frac{p}{p_o} + I \tag{6.32}$$

The linear portion of the curve is restricted to a limited portion of the isotherm, generally between 0.05–0.30 of normalized pressure. The slope and intercept are used to determine the quantity of nitrogen adsorbed in the monolayer and to calculate the surface area. For a single point method, the intercept is taken as zero or a small positive value, and the slope from the BET plot used to calculate the surface area. The surface area reported, usually quoted in $m^2 \cdot g^{-1}$, will depend upon the method used, as well as on the partial pressures at which the data were collected.

6.6.3 BET PARTICLE ANALYSIS AND THE EQUIVALENT SPHERE

To derive a particle size from BET measurements, several assumptions must be made, that is, all the particles are spherical in shape, and have sharply defined diameters and densities. Based on these assumptions an equivalent particle diameter can be calculated from the specific surface area and known density. In reality, this is not strictly valid, and so the calculated diameter is referred to as the *'equivalent sphere'* diameter, that is, an average of the particle diameters.

Additionally, further consideration should take into account the fact that the specific surface area of a particle can be divided into two parts: the outer surface and the inner surface areas. The outer surface area is correlated with the particle size and shape, whereas the inner surface area reflects the particle porosity. A distinction can be made by using gases of different molecular weight/molecular foot-print. For example, nitrogen is a relatively small molecule, with a foot-print of $0.162 \ nm^2$, and can penetrate into small pores. A full isotherm can be undertaken to determine the specific pore density as well as the specific surface area.

6.7 SURFACE AND BULK CHEMISTRY

Functionalization of nanoparticles is becoming increasingly important and common, in order to tailor the surface chemistry, and thus increase its specificity for interaction with the environment. There are numerous current and projected future applications. For instance, the mature research area of catalysis is based on the preparation of surfaces with well-defined chemistries and with the highest possible specific surface area. In the growing field of bio-nanotechnology, device functionalities based on molecular recognition are of fundamental importance. In nanocomposite structures, surface and interface functionalities are optimised in order to suppress delamination under stress, or to promote, or suppress, charge transport across phase boundaries. As micro-electromechanical systems (MEMS) make the transition to nano-electromechanical systems (NEMS), the regime surface chemistry becomes increasingly important if for no other reason than to reduce friction (in the limit, nanostructures become surfaces in contact and adhesion/friction will preclude dynamical functionality).

6.7.1 SINGLE PARTICLE ANALYSIS

The gaining of insight into the surface chemistry of a single, or a few, nanoparticle(s) calls for a spectroscopy that couples to the characteristics of electronic structure in the surface. The extent to which any such spectroscopy is practical and useful is determined by the combination of lateral spatial resolution consistent with particle diameter, surface specificity at the monolayer level, spectral resolution at the 0.1 eV level, and count rate.

There are few, if any, routine and user-friendly techniques that will provide direct, complete, and unequivocal results for surface chemistry from the analysis of single nanoparticles.

The most readily available technique is that of TEM in combination with EELS, as described in Chapter 3. While TEM/EELS has the required lateral and spectral resolution, it is not surface specific. Accordingly, the utility of the method is conditional on being able to assign particular spectral features to surface, as opposed to bulk, chemistry. The collection of reliable data requires that the particle be small, in order to ensure that signals from the surface layer dominate over those from the bulk.

If a single nanoparticle is a reasonable conductor, then STM/STS can provide atomically resolved and highly surface-specific information about surface chemistry, in combination with topography, surface reconstruction, and defect structure (see Chapter 5). However, STM/STS is not routinely available and cannot be considered to be user-friendly.

Auger microprobe spectroscopy carried out with the latest generation instrumentation offers an imaging capability comparable to that of a field-emission SEM, spectroscopic lateral resolution well below 100 nm, in combination with a surface specificity of a few monolayers. However, Auger processes are complex, leading to spectra from which chemical information can be extracted only with some difficulty. In addition, charging effects and electron beam damage are common, and constitute an additional complication.

6.7.2 SURFACE CHEMISTRY OF PARTICLE ENSEMBLES

Techniques and methods for the analysis of particle ensembles are relatively much more mature than those for the analysis of single nanoparticles, largely because particles have been industrial commodities since well before the industrial revolution. Many of those techniques have in recent times been 'reinvented' and gained additional effectiveness and user-friendliness, and have been adapted for nanoscale analysis.

X-ray photoelectron spectroscopy (XPS) has been, and remains, the most widely used technique for analysis of surfaces. The lateral resolution of current generation imaging instruments is ca. 10 μm, while the surface specificity is a few monolayers (adopting an angularly resolved configuration, the sampling depth can be limited to a single monolayer). The spectral resolution is typically 0.1–0.2 eV; the core levels of the great majority of atomic species exhibit shifts of that magnitude, or greater, with the shift being a characteristic response to the chemical environment. An XPS spectrum usually contains much additional information, such as plasmon structures and shake-up/down peaks, which constitute additional markers which can aid in interpretation. XPS has the added merit of being quantitative at the level of ± 10 at.%. Accordingly, it can provide analytical information about near-surface abundances, which constitutes a complement to analytical information from techniques that are bulk sensitive. However, XPS cannot provide analysis of individual particles, but only of ensembles, and such analysis requires that the particles present a representative surface of at least microscopic lateral extent. One way of achieving this is to embed the particles in a soft metallic foil (e.g., high purity In or Au), taking care to remove those particles that are not in direct contact with the foil.

6.7.3 WETTABILITY

In many cases, the most important attribute of a particle ensemble is its wettability (e.g., flotation processing in the mining industry). Following on from Section 6.7.2, it is obvious that wettability is a direct consequence of the surface chemistry. A hydrophilic, (or hydrophobic) surface chemistry will promote (or prevent) wetting in an aqueous fluid.

Good wetting occurs if a liquid spreads out over the substrate in a uniform film and in doing so makes a small contact angle with the substrate. Poor wetting occurs when a liquid forms droplets on the surface making a large contact angle. For a liquid to wet a surface, the liquid should have a lower surface tension, γ, than the solid's surface energy (or critical surface tension), γ_c. The two cases are illustrated in Figure 6.18.

According to Young's equation, the surface tensions (liquid/vapour: γ_{LV}, solid/liquid: γ_{SL}, and solid/vapour: γ_{SV}) at the three phase contacts are related (see Figure 6.19) to the equilibrium contact angle σ through

$$\gamma_{LV} = \gamma_{SL} + \gamma_{LV} \cos \sigma \tag{6.33}$$

For spontaneous wetting to occur, the condition must hold that

$$\gamma_{SV} > \gamma_{SL} + \gamma_{LV} \tag{6.34}$$

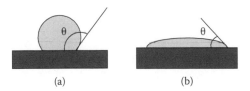

(a)   (b)

FIGURE 6.18 Figures A and B demonstrate a difference in wettability. Figure A shows how a water droplet might appear on a hydrophobic surface such as wax, while Figure B shows how it might appear on a hydrophilic surface such as a contact lens. (Note: contact angle $\theta = \sigma$.)

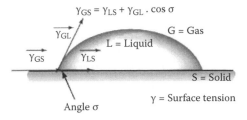

FIGURE 6.19 Angle of contact of a drop of liquid with the surface of a solid object. (Adapted from http//:en.wikipedia.org/wiki/filecontact_angle.sug.)

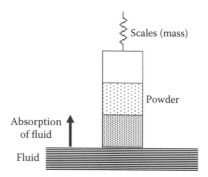

FIGURE 6.20 Schematic arrangement of the Washburn method for measuring the wettability of a powder sample.

The condition for good wetting is $\gamma_{SV} < \gamma_{LV}$.

The phenomenon of wetting has static and dynamic dimensions, as described in a seminal review (de Gennes 1985). Surface structure has been identified as a significant issue in wetting and has recently received considerable attention (Cassie and Baxter 1944; Watson and Watson 2004; Wenzel 1936).

Wettability of particle ensembles can be carried out in the same manner, by goniometric methods, as for solid surfaces, namely, by preparing a close-packed flat powder surface, and then placing a drop of fluid on the surface. In the case when the contact angle is less than 90°, spontaneous penetration will take place, and will complicate the analysis. Likewise, the inevitable surface roughness may lead to misleading dynamic results.

The Washburn method (Washburn 1921) is yet another 'traditional' technique that has found application in nanotechnology. It is based on the following relationship

$$T = \frac{\eta M^2}{C\rho^2 \gamma \cos\theta} \tag{6.35}$$

where T is time after contact, η is viscosity, ρ is density of fluid, γ is surface tension of fluid, θ contact angle and M is mass adsorbed fluid. C is a material constant, which remains fixed for exposure to different fluids. Thus, it can be eliminated as a variable by performing two or more measurements on the same powder with different fluids.

A typical experiment involves measuring, as a function of time, the mass of fluid absorbed by a porous plug that is in contact with a fluid reservoir. The arrangement is shown schematically in Figure 6.20.

6.8   OVERVIEW—CHOICE OF TECHNIQUE(S)

This chapter has addressed the question of how to characterize nanoparticles (i.e., objects on the size-scale well below that corresponding to the optical diffraction limit). The techniques fall into two broad categories, depending on what is meant by characterization.

TABLE 6.1

Characterisation of Individual Nanoparticles

Attribute(s)	Technique(s)	Comments
Crystal structure	HRTEM	Lattice imaging, see Chapter 2
Composition	EDS/EELS	(S)TEM/SEM, see Chapter 3
Molecular structure		Useful information from EELS, see Chapter 3
Size and shape	SEM/TEM/AFM	See Chapters 2 and Chapter 5
Surface chemistry		Useful information from AFM/F-d, usually in an aqueous solution, see Chapter 5
Surface area	SEM	Inferred/estimated from imaging, see Chapter 2

In an industrial context, quality control becomes an important variation on the general theme of characterization. The emphasis is then essentially on ensuring that nothing has changed from one day to the next, or from one batch to another, where 'nothing' might refer to the functionality of the product (e.g., nanoparticles with a particular desired functionality).

In an R&D context, on the other hand, the aim is to determine the attributes of a single or a small number of nanoparticles, where attributes refer to such things as crystalline structure, composition, size, morphology, and so forth. This kind of characterization usually takes place during an R&D phase, where a great deal of expertise, time, energy, and cost can be invested in one-of, or a small number of, analyses. The technique(s) can then be chosen from a category that is generally not cost-effective in the case of quality control. The relative freedom to choose may also be exercised for the purposes of fault-finding and/or problem-solving, when quality control says that there is a problem, and when quality control does not give sufficient information to solve the problem.

Another category of techniques, of greater relevance to quality control, is concerned with obtaining ensemble averages of attributes of nanoparticles, again with respect to structure, composition, size, and so forth. In many cases, it is neither practical nor necessary to carry out analyses on a sufficiently large number of individual particles in order to infer average or representative values for a large population of, say, particles from a broad size range, and possibly with different compositions, surface chemistries, and so on.

The most widely used and available techniques from the two categories have been listed in Table 6.1 and Table 6.2 with reference to information obtainable from single nanoparticles or/and ensembles of such particles.

ACKNOWLEDGEMENTS

This chapter is, in part, based on an earlier chapter authored by M. Werner, A. Crossley and C. Johnston in *Handbook of Surface and Interface Analysis*. Input to this chapter from, and critical reading by, Drs. C. Johnston and A. Crossley of Begbroke Nano at the Oxford University Begbroke Science Park is greatly appreciated.

TABLE 6.2
Characterization of Ensemble Averages

Attribute(s)	Technique(s)	Comments
Crystal structure	XRD*	Down to 10 nm particle size
Composition	XPS*/EDS*/WDS*	See Chapter 3
Molecular structure	Raman*/IR*EXAFS†*	See Chapter 4
Number density	CPC**	Down to 3 nm size range, dynamic range up to 10^6 particles/cm^3
Equivalent size	DLS	
DLS		
DSC	Monodisperse samples	
Bimodal size distributions		
Multi-modal size distributions		
Surface chemistry	XPS*/SRXPS†	—
Surface area	BET	—

* Specimen configuration, usually dense thin packed layer of particles on flat substrate with typical coverage of 1 cm^2.

** Condensation Particle Counter.

† Synchrotron-based technique.

REFERENCES

Brunauer, S., 1943, *The Adsorption of Gases and Vapours*, Oxford: Oxford University Press.

Brunauer, S., Emmett, P. H. and Teller, E., 1938, *J. Am. Chem. Soc.*, 60, 309.

Cassie, A. B. D. and Baxter, S., 1944, *Trans. Faraday Soc.*, 49, 546.

CPS Application note, CPS Instruments, www.cpsinstruments.eu/pdf/General%20Brochure.pdf.

Davis, W., 2009, *Intl. Gases Instrum.*, August 29.

De Boer, J. H., 1953, *The Dynamical Character of Adsorption*, Oxford: Clarendon Press.

De Gennes, P. G., 1985, *Rev. Mod. Phys., Part 1*, 97, 827.

Derjaguin, B. V. and Landau, E. M., 1941, *Acta Physicochim.*, URSS 14, 633.

DLS Instrumentation, Dynamic Light Scattering: An Introduction in 30 Minutes, Malvern, www.malvern.com. (Downloadable application Note.)

Doremus, R., 2002, *Langmuir*, 18, 2436.

IUPAC Report, 2005, Measurement and Interpretation of Electrokinetic Phenomena, IUPAC Technical Report, *Pure Appl. Chem.*, 77, 1753.

Malvern ZetaSizer NS, 2007, Operating Manual, Malvern, www.malvern.com. (Also see Zeta Potential: An Introduction in 30 Minutes) (Downloadable application note.)

Mie, G., 1908, *Ann. Phys.*, 25, 377.

Pecora, R. (ed.), 1985, *Dynamic Light Scattering: Applications of Photon Correlation Spectroscopy*, New York: Plenum Press.

Rawle, A., Basic Principles of Particle Size Analysis, Malvern, www.malvern.com. (Downloadable technical paper.)

Schulenburg, M., 2006, *Nanotechnologie: Innovationen für die Welt von Morgen*, Hrsg, Bonn/Berlin: BMBF.

Van de Hulst, H. C., 1957, *Light Scattering by Small Particles*, New York: Wiley.

Wikipedia.org/wiki/File:BET-1.jpg.

www.refiningonline.com/engelhardkb/crep/TCR2–10.htm.

Verwey, E. J. W. and Overbeck, J. Th.G., 1948. *Theory of the Stability of Lycophobic Colloids*, Amsterdam: Elsevier.

Watson, G. S. and Watson, J. A., 2004, *Appl. Surf. Sci.*, 235, 139.

Washburn, E. W., 1921, *Phys. Rev.*, 17, 374.

Wenzel, R. N., 1936, *Indust. Eng. Chem.*, 28, 988.

Section II

Applications

7 C$_{60}$ and Other Cage Structures

7.1 INTRODUCTION

The unequivocal identification and description of the C$_{60}$ molecule in 1985 (Kroto et al. 1985) predate the discoveries of other nanostructural building blocks such as nanotubes, nanowires, and graphene. Thus, it can said that C$_{60}$ ushered in the era when nanotechnology became a practical proposition. However, in spite of the early predictions, and unlike the subsequently discovered nanotubes, fullerenes and fullerites have not made much practical impact. Nevertheless, they have been interesting objects from the point of view of fundamental science. The first ten years of work in the field have been reviewed in the literature (e.g., Kumar, Martin, and Tosatti 1993; Eletskii and Smimov 1995).

The C$_{60}$ structure, sometimes known as *'Buckminsterfullerene'*, is shown in Figure 7.1. It consists of 20 hexagons and 12 pentagons, in which the carbon atoms are located at the vertices of a truncated icosahedron with I$_h$ symmetry.

7.1.1 EULER'S THEOREM

Nearly 300 years ago, Euler constructed a geometrical theorem for polyhedra that can describe fullerene molecules of the type C$_n$. According to Euler,

$$v + f = e + 2 \tag{7.1}$$

where v is the number of vertices, f the number of faces, and e the number of edges. In the case of a fullerene $n = v$, and

$$e = \frac{3n}{2}$$

(each vertex is the intersection of 3 edges, with the factor of 2 arising from the double-counting of edges). Then

$$f = \frac{n}{2} + 2$$

The faces are pentagons (p) and hexagons (h) with 5 and 6 vertices, respectively. Accordingly,

FIGURE 7.1   Structure of C_{60}.

$$n = \frac{5p + 6h}{3} \tag{7.2}$$

where the factor of 3 arises from the triple-counting of the vertices. Thus,

$$f = p + h = \frac{n}{2} + 2 \tag{7.3}$$

It is straightforward to solve for p and h from Equations (7.2) and (7.3), resulting in

$$h = \frac{n}{2} - 10 \tag{7.4}$$

$$p = 12$$

In the case of $C_{60,}$ the molecule will have 12 pentagonal and 20 hexagonal faces.

7.1.2   THE FULLERENE FAMILY

The original two members, C_{60} and C_{70}, of the family of carbon-based cage molecules, denoted C_n, have been joined by a large number of related structures, with n ranging from 20 up to 120, and beyond. Their structures and symmetries have been described systematically in a scheme proposed by IUPAC (Cozzi, Powell, and Thilgen 2005).

While the discovery of $C_{60/70}$ attracted interest by virtue of its properties as a unique molecular species, the focus of attention moved rapidly on to other carbon-based cage structures, and to systems where fullerenes were integral components of complex structures. An abbreviated list includes the following:

- **Fullerenes—C_n—in isolation**: The family of non-interacting C_n-type carbon-based cage structures.
- **Endohedral fullerenes**: As early as 1985, it was proposed that atomic dopants could be confined within the cages of C_n structures and thus lead to interesting and potentially useful changes in their structures and electronic properties (Kroto et al. 1985). The structures became known as 'endohedral fullerenes' with the general notation $A_x B_y @ C_n$ (allowing for multiple non-equivalent dopants).
- **Functionalized fullerenes**: The fullerene cage structures are relatively inert chemically, and are sparingly soluble in most common solvents. In order to incorporate them into larger chemical complexes, or into solids for materials applications, the molecules must be functionalised (derivatised) by one of several synthesis routes. Details of the chemistry involved are outside the scope of this volume, and the reader is referred to the review literature (e.g., Wudl 2002 and references therein).
- **Polymeric fullerene structures**: Covalent bonds can be formed between C_n molecules in a crystal by irradiation with near-ultraviolet (UV) laser light, under high hydrostatic pressure, or by doping. In general, the polymerisation is established by a cycloaddition reaction, leading to parallel oriented double bonds between neighbouring molecules. Examples of polymeric structures in which the bonding is between nearest neighbour C_{60} molecules (e.g., a four-membered ring (tp)), are illustrated in Figure 7.2.
- **Fullerites**: A polymeric solid fullerite phase prepared from molecular C_{60} at high pressure (typically >10 GPa) and temperature (typically >600K) (Blank et al. 1997) has been shown to exhibit exceptional hardness (greater than that of diamond (Blank et al. 1998)). The material is known variously as 'ultrahard fullerite', 'phase V', and 'Am2'.
- **Fullerenes as 'peapod' structures**: A carbon nanotube can be considered as a long cylindrical container with a diameter of ca. 1 nm. Accordingly, it may be viewed as a perfect reaction vessel for exploring the 1-D quantum

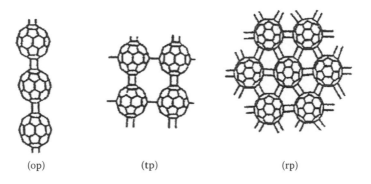

(op)   (tp)   (rp)

FIGURE 7.2 Polymeric fullerene phases: (op) one-dimensional orthorhombic; (tp) two-dimensional tetragonal; (rp) two-dimensional rhombohedral. (From Kuzmany et al., 2004, *Phil. Trans. R. Soc. Lond. A,* with permission from the Royal Society.)

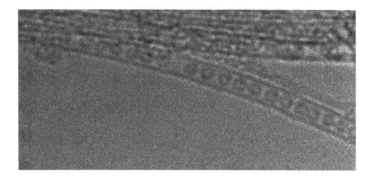

FIGURE 7.3   HRTEM image of fullerene molecules located inside a SWCNT. (Reproduced with permission from Khlobystov, 2011.)

realm of physics and chemistry with nanoscale confinement in two dimensions. The arrangement with fullerene molecules located within the core of a single wall CNT (SWCNT) has attracted particular attention, and merits the designation of 'peapod'; an example is shown Figure 7.3.

7.2   CHARACTERIZATION OF FULLERENES AND FULLERENE COMPOUNDS

Non-interacting fullerene molecules are zero-dimensional objects with diameters of ca. 1 nm. Thus, they represent a considerable challenge to the materials characterization community. Their general shape is resolvable by high resolution transmission electron microscopy (HRTEM) analysis, but details of their structure cannot be determined unequivocally from images. However, the inclusion of higher-Z atomic dopants in endohedral fullerenes can generally be confirmed by HRTEM/STEM methods. STM can provide more detailed topographical information, and some electronic information, but even the best present work has not succeeded in shedding much light on the molecular structure.

The situation is much better in the cases of fullerene solids and fullerene ensembles. Raman spectroscopy, and to a lesser extent IR spectroscopy, have been the favoured tools. The resulting spectra reflect the symmetries, and therefore the allowed vibrational modes, which, with a great deal of confidence, can be related directly to the molecular structures of isolated fullerenes, or to the structure of fullerene-based solids. As well, the allowed modes are sensitive to the presence of dopants and to interactions of fullerenes in confinement.

7.2.1   CHARACTERIZATION OF NON-INTERACTING C_{60} AND OF $C_{N \neq 60}$ MOLECULES

High performance liquid chromatography (HPLC) in combination with spectroscopic or mass spectrometric analysis can readily separate, identify, and quantify the fractions of C_{60} and C_{70} present in carbonaceous material from a synthesis route. Likewise, many of the techniques and methods relevant to nanoparticle analysis, described in Chapter 6, can be applied to bulk analysis of particle size and distribution.

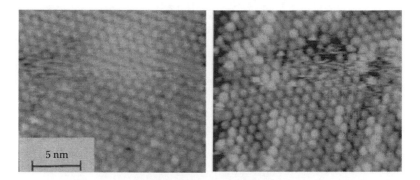

FIGURE 7.4 STM images of ordered fullerene molecules deposited onto an Au(111) substrate by sublimation from a powder source. (Left) From a pure C$_{60}$ source. (Right) From a mixed C$_{60}$/C$_{70}$ source. The spheres with brighter contrast and larger diameters represent the C$_{70}$ molecules. Note that the images are consistent with diameters of 0.7 and 0.8 nm, respectively, for C$_{60}$ and C$_{70}$. (Adapted from Lang et al., 1993, *Appl. Phys. A*, with kind permission from Springer Science+Business Media.)

Since C$_{60}$ has icosahedral I$_h$ symmetry—the highest possible molecular symmetry, its Raman spectrum is fairly simple. The most prominent feature derives from the A$_{g2}$ mode at 1470 cm^{-1}. It can be visualised as a pentagonal pinch mode. A weaker A$_{g1}$ mode arises from the radial breathing mode and gives rise to a peak at 496 cm^{-1}. While there are a total of ten Raman-active modes, the two aforementioned peaks, at 1470 and 496 cm^{-1}, are the most widely used analytical Raman signatures for the structural and electronic properties of C$_{60}$. The other fullerenes with n < 60 or > 60 have a number of isomers with more complicated symmetries. C$_{80}$, for example, has seven isomers, including one with I$_h$ symmetry (Dresselhaus, Dresselhaus, and Eklund 1996). A consequence of the reduced symmetries and the larger number of cage atoms is that the Raman spectra of these fullerene molecules will exhibit more complexity than the spectrum of pure C$_{60}$. For instance, group theory suggests that C$_{70}$ has 53 Raman-active modes and 31 infrared-active modes (Guha and Nakamoto 2005).

In the early days, characterization by STM provided useful information on the relative size distribution of a mixed C$_{60}$/C$_{70}$ powder specimen deposited on an Au(111) surface (Lang et al. 1993), see Figure 7.4. The powders were also analysed by ^{13}C NMR. Due to the equivalence of all carbon sites defining the C$_{60}$ structure, a single line was observed at 143.6 ppm (with respect to tetramethylsilane). On the other hand, C$_{70}$ has five non-equivalent carbon sites, resulting in five NMR lines. The areas under the five lines for C$_{70}$ were integrated and a ratio was determined with respect to the area under the single line for C$_{60}$, in order to arrive at an estimate for the relative abundances of the two species. The results were consistent with those obtained from analysis of the topographical STM images.

Two related STM studies (Altman and Colton 1992 and 1993) have investigated the adsorption of C$_{60}$ molecules on Au(111) and Ag(111) surfaces. They represent a good example of the merits of scanning probe techniques. The two studies were concerned with the growth of fullerene films by vapour deposition, and they determined

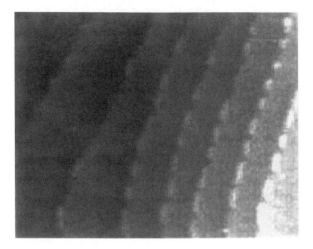

FIGURE 7.5 Illustration of preferential adsorption and growth of fullerene chains at edges of terraces at a coverage of 0.04 ML. The field of view is 70 × 58 nm², and the image was obtained with a tip bias of 1.5 V and a tunnel current of 0.5 nA. (From Altman and Colton, 1992, *Surf. Sci.,* with permission from Elsevier.)

that there was preferential adsorption and nucleation at the intersections of multiple steps and at the edges of monatomic steps on substrate terraces. Formation of arrays of short chains was observed at increasing coverage, as seen in Figure 7.5. At higher coverages, the formation of ordered monolayer structures was constrained by the underlying Au(111) structure, as seen in Figure 7.6. The second layer was found to nucleate at random locations and exhibited hexagonal island structures with a lattice constant of ca. 1.0 nm, as seen in Figure 7.7. High-resolution imaging, Figure 7.8, suggests that the orientation of the adsorbed fullerene presented a pentagon face uppermost, thus defining the orientation of the adsorbate.

7.3 ENDOHEDRAL FULLERENES

During the synthesis of fullerenes, atoms or molecules can be trapped inside the cage, resulting in so-called endohedral fullerenes (endohedral doping). This was first demonstrated with $La@C_{60}$ and $La_2@C_{60}$ (Heath et al. 1985). The early work on endohedral fullerenes focussed on rare earth and transition metal atom dopants, but subsequently compounds such as $Sc_3N@C_{80}$, $Sc_3N@C_{82}$, and $Sc_2C_2@C_{84}$ were also synthesised. The Raman and IR spectroscopy of endofullerenes has been reviewed by Krause and Kuzmany (2002). An example of the IR and Raman spectral signatures is shown in Figure 7.9.

The IR spectrum of $Sc_3N@C_{80}$ is dominated by a strong line at 597 cm⁻¹ and has a complex structure below ca. 1400 cm⁻¹. Other lines were observed at ca. 500, 1200, 1445, and 1510 cm⁻¹. The Raman spectra are more complex. A cluster of broad structures was present in the range 1000 to 1600 cm⁻¹, and multiple lines were found between 200 and 815 cm⁻¹. The strong lines at 411, 232, and 210 cm⁻¹ were present and prominent for all three excitation wavelengths. The dependence of the modes

FIGURE 7.6 An ordered fullerene structure at intermediate coverage showing order imposed by the Au(111) reconstructed substrate. A domain boundary is evident (upper left corner). The bright features are likely to be due to the presence of an admixture of C_n species with n > 60. The field of view is 23×23 nm^2, and the image was obtained at a tip bias of 2 V and a tunnel current of 0.1 nA. (From Altman and Colton, 1992, *Surf. Sci.,* with permission from Elsevier.)

between 1000 and 1600 cm^{-1} on wavelength was consistent with resonance enhancement (see Figure 7.10).

The effects of rare earth doping on the Raman signatures are shown in Figure 7.10. The similarities of the spectral features above 200 cm^{-1} suggest that the carbon cage structure remained similar for all four compounds. The bands below 200 cm^{-1} at 183, 163, 162, and 155 cm^{-1}, respectively, for Y, La, Ce, and Gd are dependent on the metal dopant. The metal to cage interaction can be described by a diatomic vibration, with a frequency, v, given by (Guha and Nakamoto 2005)

$$v = 4.12\sqrt{\frac{f}{\mu}} \text{ (in non-SI units)} \tag{7.5}$$

where f is an effective force constant (in dyne/cm), and μ is the reduced mass (in atomic weight units). A linear dependence is found between v and $\mu^{-1/2}$, allowing a value for f to be obtained, thus suggesting that the metal-to-fullerene interaction remains the same for the four dopants, and that the bonding is mainly ionic (i.e., M^{3+}-C_{82}^{3-}).

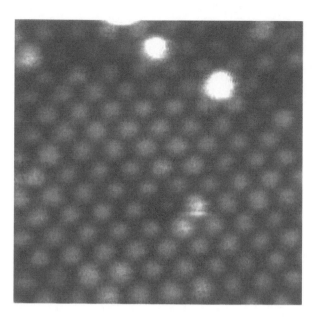

FIGURE 7.7 A second fullerene layer exhibiting hexagonal packing with a nearest neighbour distance of ca. 1 nm. The field of view was 9.4×9.4 nm^2, and the image was obtained with a tip bias of -1.5 V and a tunnel current of 0.1 nA. (From Altman and Colton, 1992, *Surf. Sci.,* with permission from Elsevier.)

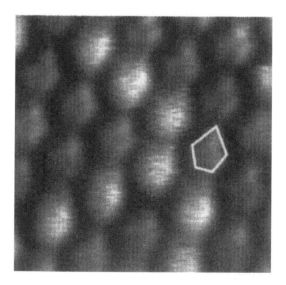

FIGURE 7.8 High resolution STM image showing the pentagonal face, viewed at an oblique angle, defining the orientation of the fullerene adsorbate. Field of view was 4.8×4.8 nm^2. (From Altman and Colton, 1993, with permission from American Physical Society.)

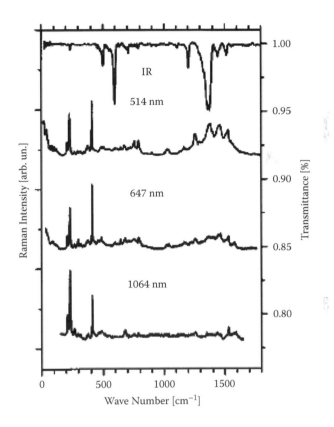

FIGURE 7.9 IR and Raman spectra for Sc$_3$N@C$_{80}$. The data were obtained at 300K. The spectra show that the choice of excitation wavelength is significant. (From Krause et al., 2001, *J. Chem. Phys.*, with permission from the American Institute of Physics.)

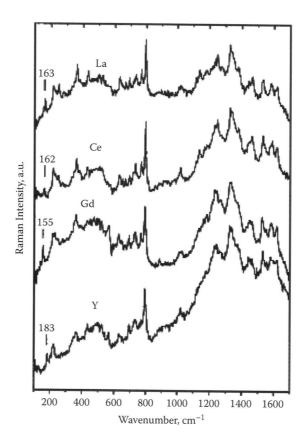

FIGURE 7.10 Raman spectra of M@C$_{82}$ rare earth doped fullerenes, with M = La, Y, Ce and Gd. (From Lebedkin et al., 1998, *Appl. Phys. A*, with kind permission from Springer Science+Business Media.)

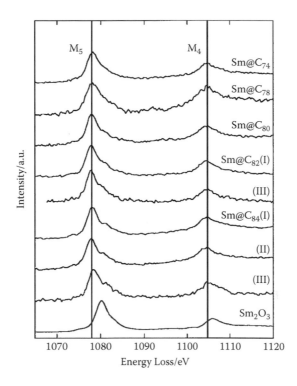

FIGURE 7.11 EELS spectra of Sm-doped endohedral fullerenes for Sm@C$_{74}$, Sm@C$_{78}$, Sm@ C$_{80}$, Sm@C$_{82}$ samples I, III, Sm@C$_{84}$ samples I, II, III, and for trivalent Sm in Sm$_2$O$_3$. (From Okazaki et al., 2000, *Chem. Phys.*, with permission from the American Institute of Physics.)

An early investigation by TEM/EELS (electron energy loss spectroscopy) of fullerene cage structures doped with Sm (Okazaki et al. 2000), used a JEOL 2010F instrument operated at 120 keV. The specimen consisted of the placement of a droplet of fullerenes in solution onto a holey carbon microgrid. The spectra covering the M$_4$ and M$_5$ regions for Sm for C$_n$, with n = 74, 78, 80, 82, 84, are shown in Figure 7.11. A reference spectrum obtained from a Sm$_2$O$_3$ specimen is also shown. Within the experimental scatter, the EELS spectra from the doped cage structures were identical, suggesting that the formal oxidation state of the Sm dopant was 2+, irrespective of the size of the cage, (as opposed to 3+ for the oxide spectrum), based on comparison with X-ray absorption spectroscopy (XAS) data in the literature (Thole et al. 1985).

7.4 FULLERITES

Fullerites may be defined as molecular solids in which the molecular constituents are fullerenes. Depending on the processing conditions, the solids may be crystalline, polycrystalline, polymeric, or amorphous, and they may be doped interstitially. Accordingly, it would be expected that the character of their vibrational spectra would arise from molecular structure (i.e., internal modes) and from intermolecular interactions (i.e., external modes). The latter point is illustrated by the Raman spectra

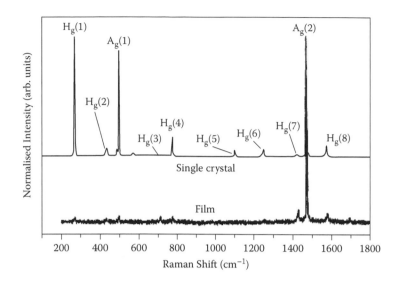

FIGURE 7.12   Raman spectrum of a C_{60} single crystal (1064 nm excitation) and a polycrystalline C_{60} film (488.0 nm excitation) recorded at 90 K. The spectra were normalised to a total height of unity. (From Kuzmany et al., 2004, *Phil. Trans. R. Soc. Lond. A,* with permission from the Royal Society.)

in Figure 7.12. The dominant feature in the spectrum for an adsorbed film of non-interacting C_{60} molecules is the peak corresponding to the characteristic A_{g2} mode, at 1469 cm^{-1}. That peak is present as a strong feature in the spectrum for the crystalline specimen, but there are numerous other lines, (labelled with Mullikan indices), that arise from the breaking of the symmetries that account for forbidden modes for isolated fullerene molecules.

Much of the early work on fullerite after its discovery was concerned with doping with alkali metals in order to tailor the electronic structure. (The result was metallic conduction and, in some cases, superconductivity.) Representative Raman spectra are shown in Figure 7.13.

The effect of doping fullerite with an alkali, on the A_{g2} mode, is shown in Figure 7.14.

The AC_{60} phase, where A is an alkali, exhibits an face-centered cubic (FCC) structure at high temperature, with the principal fullerene mode peak shifting from 1468 down to 1461 cm^{-1}. During slow cooling, the KC_{60} undergoes phase separation at 420K to C_{60} and $K_{3}C_{60}$. On the other hand, RbC_{60} and CsC_{60} have phase transitions at a temperature similar to that of an orthorhombic polymeric state, in the form of a linear chain. Finally, if the KC_{60} high temperature phase is quenched through the phase-separation temperature, it also transforms into an orthorhombic polymeric phase. This behaviour is shown for the two phases KC_{60} and RbC_{60} in Figure 7.15. The Raman response of the principal fullerene mode is taken to be an indicator of the structural changes. For the potassium compound, the splitting of the line originally at 1461 cm^{-1} for the KC_{60} FCC phase into two lines at 1468 cm^{-1} and 1448 cm^{-1} is apparent. RbC_{60} remains stable until a temperature of ca. 397 K is reached. At 380

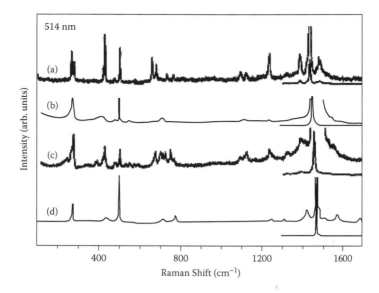

FIGURE 7.13 Raman spectra for K_xC_{60} fullerites. (a) K_6C_{60}, and (b) K_3C_{60} at room temperature, and (c) K_3C_{60} at low temperature. The spectrum in (d) for undoped C_{60} is shown for comparison. (From Kuzmany et al., 2004, *Phil. Trans. R. Soc. Lond. A*, with permission from the Royal Society.)

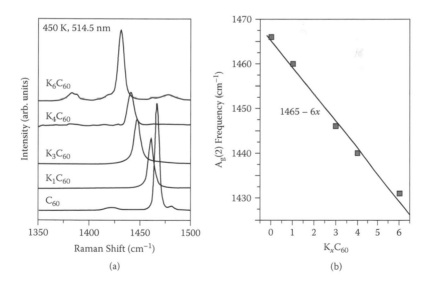

FIGURE 7.14 (a) Dependence of the $A_{g2}C_{60}$ fullerene mode peak in interstitially doped $K_x(C_{60})^{-x}$ (where x refers to the formal charge transfer between dopant and cage structure) fullerite specimens. The spectrum for C_{60} is shown for comparison. (b) An approximate linear dependence is evident for the shift of the A_{g2} frequency with increased charge transfer to the C_{60} cage. (From Kuzmany et al., 2004, *Phil. Trans. R. Soc. Lond. A*, with permission from the Royal Society.)

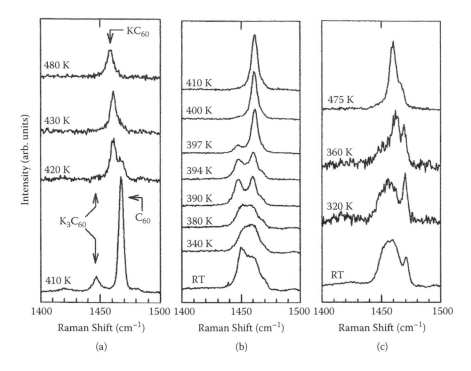

FIGURE 7.15 Temperature dependence of the Raman response for the principal fullerene mode, on cooling from the high-temperature FCC phase to the room-temperature phases. (a) Phase separation for KC_{60}; (b) polymerisation of RbC_{60}; (c) quench leading to polymerization of KC_{60}. The sharp peak at the right edge of the spectrum in (c) represents an undoped C_{60} part of the sample. (From Kuzmany et al., 2004, *Phil. Trans. R. Soc. Lond. A,* with permission from the Royal Society.)

K, the original single peak has become a double-peak structure, the latter being a characteristic of fullerene polymeric phases. A similar evolution of the KC_{60} phase is seen at ca. 320K during the quench.

Fullerite films synthesised at high temperature and pressure have attracted attention due to early reports of exceptional hardness, called '*superhard fullerite*' (Blank et al. 1997, 1998). A great deal of work was then carried out in order to establish the structure of the material, and to define optimum synthesis conditions. *In situ* Raman analyses have proven particularly useful. An example of *in situ* Raman results is shown in Figure 7.16.

The spectrum recorded at 22.5 GPa shows two broad asymmetric peaks centred at ca. 700 and 1600 cm^{-1}, respectively. After quenching, the sample exhibits a spectrum with a number of well-resolved peaks, which has some similarities to that of C_{60}, but exhibits a number of additional peaks below 800 cm^{-1}. As well, the A_{g2} mode is shifted to 1464 cm^{-1}, some 5 cm^{-1} below the value for C_{60} molecules. A comparison with the literature shows that the spectrum of the quenched sample was nearly identical to that obtained from orthorhombic C_{60}. The phase has been synthesised by Rao et al. (1997) and a polymeric structure was reported in which the C_{60} cages are joined

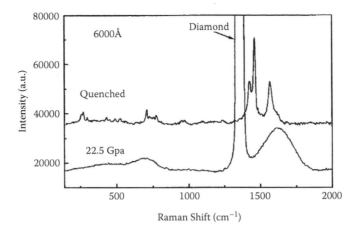

FIGURE 7.16   Raman spectra for C$_{60}$ fullerite films of 600 nm thickness obtained *in situ* at 22.5 GPa pressure, and after quenching to room temperature conditions. The prominent mode peak labelled 'Diamond' arises from the anvil of the press. (From Talyzin et al., 2001, *Phys. Rev. B*, with permission from the American Physical Society.)

by square rings to form one-dimensional chains by a 2-2 cycloaddition mechanism. A characteristic of the polymeric structure is the broad feature at ca. 900–1000 cm^{-1}. Accordingly, the data in Figure 7.16 suggest that the high-pressure conditions do not destroy the C$_{60}$ cages, and also show that 'superhard' fullerite can be transformed back to the well-known 1-D structures. Processing variables, such as film thickness and pressure, have also been investigated (Talyzin, Dubrovinsky, and Jansson 2001).

7.5   PEAPOD—FULLERENES IN CNT

The filling of CNTs by a great variety of atomic and molecular species dates back to 1998. The history, discoveries, potential applications, and progress have been reviewed (Monthioux 2002); more recently, the infusion of fullerene molecules into SWCNTs (C$_n$@SWNT is a common notation) has also been reviewed, with an emphasis on theory (Krive, Shekhter, and Jonson, 2006).

The HRTEM image in Figure 7.17 shows the structure of an SWCNT filled with KI.

The structure of K-doped fullerene peapods, (potassium is an n-type dopant), has been investigated by HRTEM and EELS. Somewhat unexpectedly, it was found (Guan et al. 2005) that the potassium dopant atoms occupied intermolecular sites between C$_{60}$ peapods. The EELS spectrum of the L edge revealed clearly the presence of potassium in the peapod. A series of images is shown in Figure 7.18, and the EELS spectra are shown in Figure 7.19.

As would be expected, the Raman spectrum of a peapod structure will contain structure from the fullerene molecule, as well as from the CNT. Additional lines and shifts will then reflect interactions, such as those of fullerene-to-fullerene and fullerene-to-CNT. The spectra in Figure 7.20 illustrate the situation.

The spectra in Figure 7.20 show that a weighted sum of the C$_{60}$ film and empty SWCNT modes can account for the main features in the peapod spectrum. The A$_{g2}$

spectrum of the C_{60} peas is shifted down by ca. 6 cm^{-1} and appears with its main peak at 1463 cm^{-1} (Figure 7.21).

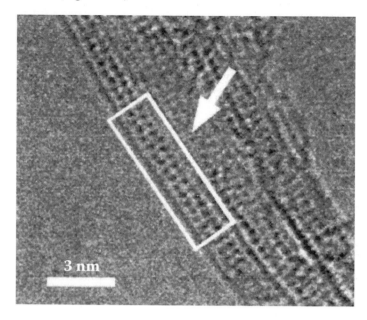

FIGURE 7.17   HRTEM image of the core of an SWNT filled with KI (KI@SWNT). A crystalline KI phase with a width of two atoms has nucleated within an SWNT of 1.4 nm diameter. Each dark dot corresponds to two superimposed atoms (i.e., either K–I or I–K), The spacing between dots along the axis of the SWNT is about 0.35 nm, that is, comparable to the (200) d-spacing in a bulk KI single crystal. The spacing between dots in the transverse direction is about 0.4 nm. (From Sloan et al., 2000, *Chem. Phys. Lett.*, with permission from Elsevier.)

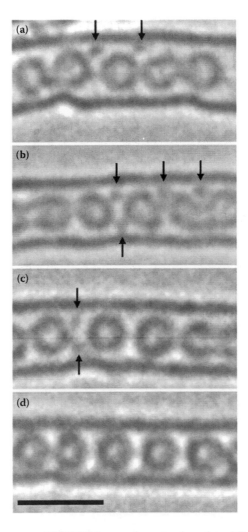

FIGURE 7.18 A sequence of HRTEM images for potassium-doped C_{60} peapods (a)–(c) and an undoped C_{60} peapod (d). The arrows indicate the potassium atom positions. Scale bar represents 2 nm. (From Guan et al., 2005, *Phys. Rev. Lett.*, with permission from the American Physical Society.)

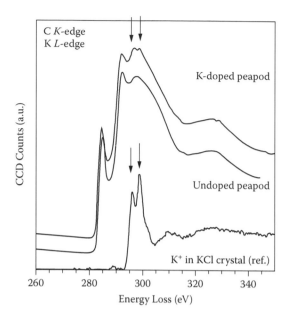

FIGURE 7.19 TEM/EELS spectra recorded from K-doped (top curve) and undoped (middle curve) C_{60} peapods within SWCNTs. The spectrum from crystalline KCl is also shown as a reference (lower curve). (From Guan et al., 2005, *Phys. Rev. Lett.*, with permission from the American Physical Society.)

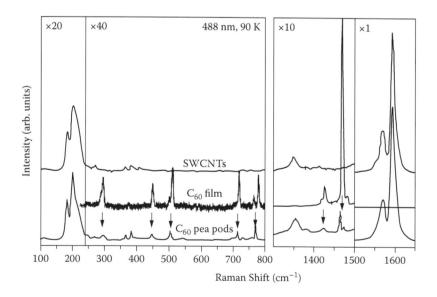

FIGURE 7.20 Raman spectra recorded using 488 nm excitation, of empty reference SWCNTs (top), a polycrystalline C_{60} film (middle), and C_{60} peapods at 90 K. Arrows indicate the response from the peas. (From Kuzmany et al. 2004, *Phil. Trans. R. Soc. Lond. A,* with permission from the Royal Society.)

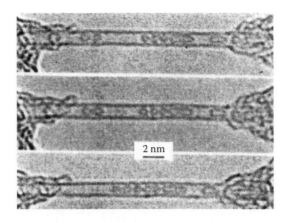

FIGURE 7.21   High resolution TEM images of (top) an example of a fullerene in which C$_{120}$ and C$_{180}$ have entered the CNT core as elongated capsules in the section to the right of the CNT, together with regular C$_{60}$ molecules to the left, within an SWCNT of diameter 1.4 nm, and of (top to bottom) the diffusion of C$_{60}$ chains (a doublet in this example) along the cavity of the SWNT due to the 100 keV electron irradiation within the microscope. The time between each image of the sequence was ca. 10 s. (From Monthioux, 2002, *Carbon*, with permission from Elsevier.)

REFERENCES

Altman, E. J. and Colton, R. J., 1992, *Surf. Sci.* 279, 49.
———Altman, E. J. and Colton, R. J., 1993., *Phys. Rev. B*, 24, 18244.
Blank, V., Buga, S., Serebryanaya. N., Dubitsky, G., Bagramov, R. Popov, M., Prokhorov, V., and Sulyanov, S., 1997, *Appl. Phys. A,* 64, 247.
Blank, V., Popov, M., Pivovarov, G., Lvova, N., Gogolinsky, K., and Reshetov, V., 1998, *Diamond Relat. Materials* 7, 427.
Cozzi, F., Powell, W. H., and Thilgen, C., 2005, (Numbering of Fullerenes, IUPAC Recommendation), *Pure Appl. Chem.* 77, 843.
Dresselhaus, M. S., Dresselhaus, G., and Eklund, P. C., 1996, *Science of Fullerenes and Carbon Nanotubes*, New York: Academic Press.
Eletskii, A. V. and Smirnov, B. M., 1995, *Usp. Fiz. Nauk.*, 169, 977.
Guha, S. and Nakamoto, K., 2005, *Coord. Chem. Rev.*, 249, 1111.
Guan, L., Suenaga, K., Shi, Z., Gu, Z., and Iijima, S., 2005, *Phys. Rev. Lett.,* 94, 045502.
Heath, J. R., O'Brien, S. C., Zhang, Q., Liu, Y., Curl, R. F., Tittel, F. K., and Smalley, R. E., 1985, *J. Am. Chem. Soc.,* 107, 7779.
Khlobystov, A., 2011, private communication.
Krause, M. and Kuzmany, H., 2002, Raman and Infrared Spectra of Endohedral Fullerenes, In *Endo-Fullerenes: A New Family of Carbon Clusters*, Akasaka, T. and Nagase, S. (eds.), p. 170, Deventer, Holland: Kluwer.
Krause, M., Kuzmany, H., Georgi, P., Dunsch, L., Vietze, K., and Seifert, G., 2001, *J. Chem. Phys.* 115, 6596.
Krive, I. V., Shekhter, R. I., and Jonson, M., 2006, *Low Temperature Physics,* 32, 887.
Kroto, H. W., Heath, J. R., O'Brien, S. C., Curl, R. F., and Smalley, R. E., 1985, *Nature,* 318, 162.
Kumar, V., Martin, T. P., and Tosatti, E. (eds.), 1993, *Clusters and Fullerenes*, Singapore: World Scientific Publishing Co.
Kuzmany, H., Pfeiffer, R., Hulman, M., and Kramberger, C., 2004, *Phil. Trans. R. Soc. Lond. A,* 362, 2375.

Lang, H. P., Thommen-Geiser, V., Bolm, C., Felder, M., Frommer, J., Wiesendanger, R., Werner, H., Schlögl, R., Zahab, A., Bernier, P., Gerth, G., Anselmetti, D., and Güntherodt, H. J., 1993, *Appl. Phys. A*, 56, 197.

Lebedkin, S., Renker, B., Heid, R., Schober, H., and Rietschel, H., 1998, *Appl. Phys. A*, 66, 273.

Monthioux, M., 2002, *Carbon*, 40, 1809.

Okazaki, T, Suenaga, K., Lian, Y., Gu, Z., and Shinohara, H., 2000, *Chem. Phys.*, 113, 9593.

Rao, A. M., Eklund, P. C., Hodeau, J. L., Marques, L., and Nunez-Regueiro, M., 1997, *Phys. Rev. B*, 55, 4766.

Sloan, J., Novotny, M. C., Bailey, S. R., Brown, G., Xu, C. Williams, C., Friedrichs, V. C., and Hutchison, J. L., 2000, *Chem. Phys. Lett.*, 329, 61.

Talyzin,, A. V., Dubrovinsky, L. S., and Jansson, U., 2001, *Phys. Rev. B*, 64, 113408.

Thole, B. T., van der Laan, G., Fuggle, J. C., Sawatzky, G. A., Karnatak, R. C., and Esteva, J. M., 1985, *Phys. Rev. B*, 32, 5107.

Wudl, F., 2002, *J. Mater. Chem.*, 12, 1959.

8 Quantum Dots and Related Structures

8.1 INTRODUCTION

A quantum dot can be defined as any solid material in the form of a particle with a diameter comparable to the wavelength of an electron. The de Broglie wavelength, λ_B, of a particle, with mass m, in motion, is given by

$$\lambda_B \approx \frac{\hbar}{p} \tag{8.1}$$

where $\hbar$ is Planck's constant, and p is the linear momentum of the particle. From the Pauli uncertainty principle, the uncertainty in the linear momentum, Δp_x, of the particle is given by

$$\Delta p_x \approx \frac{\hbar}{\Delta x} \tag{8.2}$$

where Δx is the positional uncertainty. Confinement of the particle gives rise to an additional kinetic energy, E_{conf}, of

$$E_{conf} = \frac{(\Delta p_x)^2}{2m} \approx \frac{\hbar^2}{2m(\Delta x)^2} \tag{8.3}$$

The additional energy of confinement will be significant if it is greater than the energy of the thermal motion of the particle, that is, if

$$E_{conf} \geq \frac{1}{2} k_B T \tag{8.4}$$

where k_B is the Boltzman constant. Thus, size quantization will occur if the range of confinement, Δx, is comparable to the de Broglie wavelength of the particle. For instance, the effective mass, m^*, of an electron in a semiconductor is typically $0.1m_0$ (where m_0 is the free electron rest mass), which corresponds to $\Delta x \approx 5$ nm. It is thus possible to think of a quantum dot as a rather large atom, with the consequence that atom-like, rather than band-like, electronic structure, should be found. The example is for confinement in one dimension, where the other two dimensions may be macroscopic, but the argument can be extended readily to 2-D or 3-D confinement.

8.2 PARTICLES IN 2-D AND 3-D CONFINEMENT

Additional qualitative insight can be gained by considering the situations of particles in potential wells of either infinite or finite depth. In the case of infinite depth, there is a straightforward analytical solution to the quantum mechanical problem (e.g., Eisberg and Resnick 1974).

In 1-D confinement, the potential $V(x)$ for a symmetric well of width a, and of infinite depth, is

$$V(x) = \infty \quad \text{for } x < -\frac{a}{2} \quad \text{and} \quad x > \frac{a}{2}$$

$$0 \quad \text{for } -\frac{a}{2} < x < \frac{a}{2}$$

(8.5)

The familiar time-dependent Schrödinger equation is

$$i\hbar \frac{\partial}{\partial t} \psi(x,t) = -\frac{\hbar^2}{2m} \frac{\partial^2}{\partial x^2} \psi(x,t) + V(x)\psi(x,t)$$

(8.6)

The wave function can be written as

$$\psi(x,t) = \left[A \sin kx + B \cos kx \right] e^{-i\omega t}$$

(8.7)

where A and B are amplitude factors to be determined. Inside the region of confinement, the potential is zero, and it can be deduced that the energy E is

$$E = \hbar\omega = \frac{\hbar^2 k^2}{2m}$$

(8.8)

where

$$k = \frac{\sqrt{2mE}}{\hbar}$$

The walls at the boundary are infinitely high; therefore, the wave function must go to zero at those points, and must be of the form

$$\psi_n(x,t) = \left[A \sin k_n x + B \cos k_n x \right] e^{-i\omega_n t} \quad \text{for } -\frac{a}{2} < x < \frac{a}{2} \text{ and zero elsewhere} \quad (8.9)$$

In order to have nodes at the boundaries, the wave vector must be constrained to be of the form

$$k_n = \frac{n\pi}{a} \quad \text{where } n = 1,2,3,4, \text{ and so forth}$$

(8.10)

From normalisation, by evaluation of

$$\int_{-\infty}^{\infty} \psi * \psi \, dx = 1$$

it is obvious that $A = B = (2/a)^{1/2}$, so that the integral of the probability density over the region of confinement is unity.

The allowed energies from Equation (8.8) are then given by

$$E_n = \frac{\pi^2 \hbar^2 n^2}{2ma^2} \quad \text{for } n = 1,2,3,4, \text{ and so forth} \tag{8.11}$$

Only certain values of the total energy are allowed, that is, the system is quantised. The first few values of the energy levels are shown in Figure 8.1.

The discussion can be extended to wells of higher dimensions. A 3-D rectangular box of dimensions $a \times b \times c$ will have energy levels corresponding to k – vectors given by

$$k_{n_x,n_y,n_z} = k_{n_x}\hat{\mathbf{x}} + k_{n_y}\hat{\mathbf{y}} + k_{n_z}\hat{\mathbf{z}} = \frac{\pi n_x}{a}\hat{\mathbf{x}} + \frac{\pi n_y}{b}\hat{\mathbf{y}} + \frac{\pi n_z}{c}\hat{\mathbf{z}} \tag{8.12}$$

where $\hat{\mathbf{x}},\hat{\mathbf{y}},\hat{\mathbf{z}}$ are unit vectors along the respective cartesian axes, and k_{ni} are the components of the quantized k – vectors.

The corresponding energies are

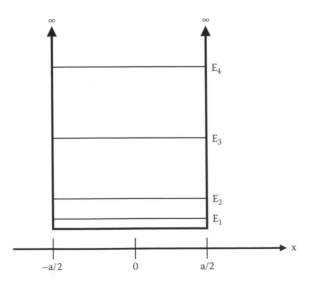

FIGURE 8.1   The energy levels of a particle confined in a 1-D infinite potential well.

$$E_{n_x,n_y,n_z} = \frac{\hbar^2 k_{n_x,n_y,n_z} \cdot k_{n_x,n_y,n_z}}{2m} = \frac{\pi^2 \hbar^2}{2m}\left(\frac{n_x^2}{a^2} + \frac{n_y^2}{b^2} + \frac{n_z^2}{c^2}\right) \qquad (8.13)$$

If two, or more, sides of the confining box are of equal length, then identical energies will be associated with multiple wave functions and degeneracy will be present in the system.

When the depth of the potential well is finite, of depth V_0, as is the case for quantum dots, some interesting new quantum mechanical features become apparent, but at the expense of a slightly more complicated solution (the situation is the same as shown in Figure 8.1, with the exception that the depth of the well is finite). The problem cannot be solved analytically, and requires a graphical method. The derivation is relatively standard (e.g., Griffiths 2005) and is limited to the symmetric 1-D case. Subscripts 0, 1, and 2 designate the regions inside the well ($V = 0$), and the left and right regions along the 1-D axis outside the well (where V is finite).

Inside the potential well, the potential is zero, and the Schrödinger equation can be written as

$$\frac{d^2\psi_0}{dx^2} = k_0^2 \psi_0 \qquad (8.14)$$

where

$$k_0 = \frac{\sqrt{2mE}}{\hbar}$$

The solution is of the form (neglecting time dependence)

$$\psi_0 = A \sin kx + B \cos kx \qquad (8.15)$$

Outside the potential well, there will be two possible forms of the Schrödinger equation. One describes the case where the solutions refer to bound states ($E < V_0$), and the other to unbound states ($E > V_0$). The former case is relevant to quantum dots. In general, the Schrödinger equation for the *unbound states* on the left (1) and right (2) sides of the well is

$$\frac{d^2\psi_{1,2}}{dx^2} = -k_{1,2}^2 \psi_{1,2} \quad \text{with} \quad k_{1,2} = \frac{\sqrt{2m(E-V_0)}}{\hbar} \qquad (8.16)$$

The general solution $\psi_{1,2}$ is of the same form as inside the well, that is, Equation (8.9). The corresponding expressions for the bound states are

$$\frac{d^2\psi_{1,2}}{dx^2} = K_{1,2}^2 \psi_{1,2} \quad \text{with} \quad K_{1,2} = \frac{\sqrt{2m(V_0-E)}}{\hbar} \qquad (8.17)$$

The general solution in this case is of exponential form.

$$\psi_{1,2} = C_{1,2}e^{-K_{1,2}x} + D_{1,2}e^{K_{1,2}x} \tag{8.18}$$

The coefficients A, B, C, D can be determined by considering various conditions, such as those at the boundaries.

The integrals over the wave functions in the range $\infty < x < +\infty$, must be of finite value. Consequently, the wave functions must be bounded and C_1 and D_2 must be zero. The overall wave functions must be continuous and differentiable over the entire range of x, and more specifically at the boundaries of the potential well. The following conditions must therefore be satisfied

$$\psi_1\left(-\frac{a}{2}\right) = \psi_0\left(-\frac{a}{2}\right)$$

$$\psi_0\left(+\frac{a}{2}\right) = \psi_2\left(+\frac{a}{2}\right) \tag{8.19}$$

$$\frac{d\psi_1}{dx}\left(-\frac{a}{2}\right) = \frac{d\psi_0}{dx}\left(-\frac{a}{2}\right)$$

$$\frac{d\psi_0}{dx}\left(+\frac{a}{2}\right) = \frac{d\psi_2}{dx}\left(+\frac{a}{2}\right)$$

Elementary manipulation of Equations (8.19) shows that there will be two solutions. One solution, with $A = 0$ and $C_1 = D_2$, is symmetric. The second solution with $B = 0$ and $C_1 = -D_2$, is anti-symmetric.

Additional manipulation gives for the symmetric case

$$C_2e^{-Ka/2} = B\cos\left(\frac{ka}{2}\right)$$

$$-KC_2e^{-Ka/2} = -kB\sin\left(\frac{ka}{2}\right) \tag{8.20}$$

whose ratio leads to

$$K = k\tan\left(\frac{ka}{2}\right) \tag{8.21}$$

The corresponding anti-symmetric expression is

$$K = -k\cot\left(\frac{ka}{2}\right) \tag{8.22}$$

The wave vectors K and k are related directly to the energy. The conditions of continuity imposed above cannot be satisfied for arbitrary values of K and k, as shown by Equations (8.21) and (8.22), and therefore not for arbitrary values of the energy either. The energies for the bound states must be quantized. The two equations cannot be solved analytically, but solutions can be found by graphical, that is, numerical, methods.

It is now convenient to define dimensionless variables and simpler notation. From the definitions of K and k, the following relationships can be written

$$u = \frac{Ka}{2},$$

$$v = \frac{ka}{2},$$

$$u^2 = u_0^2 - v^2,$$

$$u_0^2 = \frac{ma^2 V_0}{2\hbar^2}$$

(8.23)

Equations (8.21) and (8.22) can now be written in a more compact form

$$\sqrt{u_0^2 - v^2} = v \tan v = -v \cot v$$

(8.24)

for the symmetric and anti-symmetric cases, respectively.

The graphical solutions are shown in Figure 8.2. The solutions are shown as circled intercepts for the case where $u_0^2 = 20$. The corresponding energies are

$$E_n = \frac{2\hbar^2 v_n^2}{ma^2}$$

(8.25)

where n is an integer label for the quantised energies. The solution to the infinite square well can be found by letting V_0 go to infinity.

The case of the finite well illustrates two important points. The number of bound states depends on the depth of the well, but is never less than one, and there is a finite probability of finding a particle within the region of the potential barrier.

It is illustrative to consider the general changes in electronic structure and energy levels with transitions in confinement from bulk 3-D, through 2-D (thin film) and 1-D (quantum wire), to 0-D (quantum dots). This is shown schematically in Figure 8.3.

The discussion of the electronic structures of quantum wells provides a qualitative and intuitive insight into the novel physics of quantum dots. However, a complete and quantitative appreciation of spatially confined fermion systems requires more sophisticated models and mathematical methods. These are beyond the scope of this monograph, and the reader is referred to the review literature. For instance,

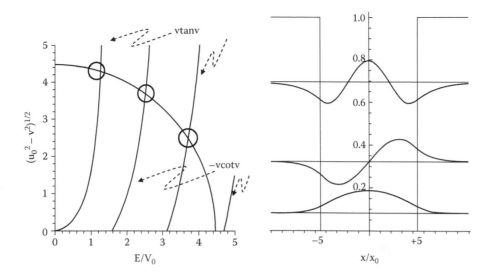

FIGURE 8.2 (Left) Illustration of graphical method, where the three circled intercepts denote the solutions $v_1 = 1.28$, $v_2 = 2.54$ and $v_3 = 3.73$. (Right) The wave functions corresponding to the three solutions. (Adapted from http://en.wikipedia.org/wiki/Finite_potential_well.)

the properties of quasi-two-dimensional semiconductor quantum dots have been discussed (Reimann and Manninen 2002, and references therein). One of the noteworthy early observations revealed an electronic shell structure, in analogy with atomic and nuclear physics, thus suggesting that aspects of earlier models, and computational techniques were transferable to quantum dots. (Kouwenhoven, Austing, and Tarucha 2001). The electronic structure of a metallic quantum dot of finite potential depth with cylindrical symmetry has been investigated with a method based on density functional theory (Lindberg and Hellsing 2005). The physics and chemistry, as well as electronic structure and transport, of semiconductor nanocrystal solids assembled from elements in the groups II, III, IV, V, VI elements, for example, CdX and PbX, where X is S, Se, Te, have been reviewed (Vanmaekelbergh. and Liljeroth 2005). Modelling with the tight-binding (Delerue and Lannoo 2004) and pseudo-potential (Wang and Zunger 1996) methods has been carried out for the single-electron wave functions and corresponding energies.

8.3 SYNTHESIS ROUTES FOR QUANTUM DOTS

There are many methods for producing quantum dots, and they can be grouped into two categories. The first group is based on solution-phase nucleation and growth, and is arguably producing products that are further along the road to commercial importance than those in the second group, principally in the biomedical industries (Jamieson et al. 2007). The second group is based on the formation of quantum well structures on substrates, by lithographic and/or epitaxial growth mechanisms. The products so produced are currently in the basic or strategic research phase, where the potential for device applications is being explored.

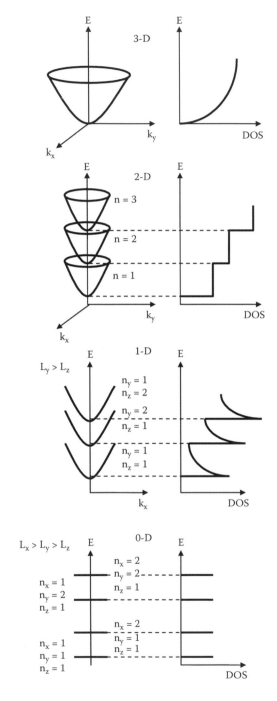

FIGURE 8.3 Schematic illustration of the transition from a 3-D band structure in a bulk solid to atomic-like structures with dimensionalities that equal the number of nanoscale dimensions. (L_i is the dimension of the confining region.)

8.3.1 COLLOIDAL NUCLEATION AND GROWTH

The earliest, and still most widely used, method for growing monodisperse quantum dots, for example, CdSe and CdS, is via a wet chemical route. Generally, it is a two-step process in which nucleation is first carried out, followed by a growth phase (Murray, Kagan, and Bawendi 2000; Rogach et al. 2002). Appropriate reagents are injected into a reaction vessel in order to establish supersaturation, and thus initiate nucleation. Nucleation will consume reagents and thus bring the system back to equilibrium. In some cases, the nucleation phase can also be controlled by temperature and/or pH change. In the second phase the remaining precursors are consumed by the growth of the particles. The nucleation phase is the critical step for ensuring a narrow size distribution. There are two methods for achieving convergence to a narrow size distribution. If the concentration of reactants is high, but below the nucleation threshold, the small particles will grow faster than particles at the higher end of the size distribution. Alternatively, at low concentration of reactants, the system will undergo a second growth phase known as *'Ostwald ripening'* in which the higher surface energy of the small particles promotes their dissolution, with redeposition onto the larger particles. The process is illustrated in Figure 8.4.

The initial precursors are generally organometallic compounds. For example, in the case of monodisperse CdSe, a mixture of elemental selenium and dimethylcadmium dissolved in trioctylphosphine is injected into a mixture of hot (ca. 300°C) trioctylphosphine oxide and hexadecylamine. Temperature, concentrations of the reactants, and the reaction time, are critical variables (Murray, Norris, and Bawendi 1993). The nucleation and growth must be carried out in an inert atmosphere, that is,

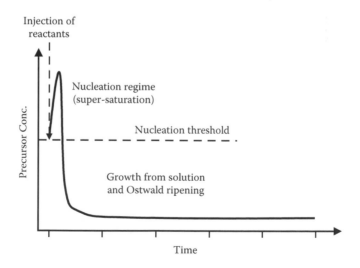

FIGURE 8.4 Schematic depiction of the nucleation and growth of monodisperse quantum dots. (The process is known as the La Mer model). The dots increase in size as a function of time and different size fractions can be extracted as the growth phase proceeds. The time scale is typically some minutes. For additional details, see Murray et al., 1993, and Murray et al., 2000.)

from which water and oxygen are excluded. Addition of a surfactant ensures that the rate of growth is sufficiently slow to allow annealing to take place, thus improving crystallinity of the end product. As well, aggregation is suppressed and surface states are passivated.

Additional processing can be carried out in order to form core-shell structures, and/or to functionalize the quantum dots for biomolecular or biomedical applications (e.g., Michalet et al. 2005, and references therein).

8.3.2   QUANTUM DOTS—LITHOGRAPHIC METHODS

High spatial resolution lithography in combination with epitaxial growth techniques accounts for a family of technologies that are being used to investigate the physics of quantum dots, with the ultimate aim of producing viable devices. There are numerous variations on this theme, and it is beyond the scope of this chapter to give a full account of the activities within the field.

A layered cylindrically symmetric arrangement can be used to define a quantum well, of some tens of nm in thickness, sandwiched between insulating layers. The device is constructed on a semiconducting substrate (e.g., n-GaAs), and the final configuration is that of a pillar of diameter some hundreds of nm.

8.3.3   SELF-ASSEMBLED QUANTUM DOTS ON A PLANAR SUBSTRATE

The growth and properties of semiconductor 3-D quantum dots have been studied extensively in the last decade. A great deal of work has been devoted to controlling the size, shape, position, and density. Much of the work has been focussed on the $(Si_{1-x}Ge_x)$-on-Si system, where a transition from planar 2-D heteroepitaxial growth to a 3-D island structure (Mo et al. 1990; Eaglesham and Cerullo 1990) has been observed. The formation of the structures is generally described in terms of the Stranski–Krastanow model, where a 2-D wetting layer of ca. 3 monolayers is followed by 3-D growth. (Brunner 2002; Baribeau et al. 2005).

8.4   CHARACTERIZATION OF QUANTUM DOTS

The functionality and topicality of quantum dots reside in their electronic properties. Being of nanoscale dimensions is a necessary condition, in general, rather than an end in itself. However, since they are of nanoscale dimensions it could reasonably be expected that those techniques and methods that are used in the characterization of particles would be relevant to the characterization of quantum dots. This is indeed so, and the reader is referred to relevant sections in Chapter 6. Here the emphasis will, therefore, be on those techniques and methods that are particularly useful in the characterization of the electronic properties of quantum dots.

8.4.1   STRUCTURE, TOPOGRAPHY, AND ANALYTICAL INFORMATION

High resolution TEM is an invaluable tool for structural analysis of individual quantum dots, as shown in Figure 8.5.

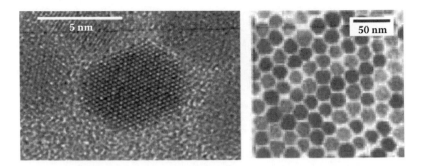

FIGURE 8.5 HRTEM of CdSe nanocrystal quantum dots. (Adapted from Manna et al., *J. Cluster Sci.*, 2002.)

Topographic imaging by AFM can be a useful and convenient tool for the study of self-assembled quantum dots on flat substrates. An example is shown in Figure 8.6. An early investigation of quantum dot superlattices made extensive use of AFM imaging (Tersoff, Teichert, and Lagally 1996). Similar results, at higher spatial resolution, but at the expense of the need to operate in an UHV environment, can be obtained by STM imaging, Figure 8.7. (e.g., Tsukamoto et. al 2006; Williams et al. 1999).

Routine EDS can at best be a convenient finger-printing tool to show that the expected elements are indeed present, as shown in Figure 8.8. High spatial resolution STEM EDS can do rather better; see Chapter 3.

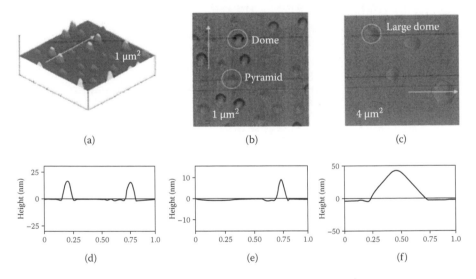

FIGURE 8.6 AFM images of the surface topography of Ge islands on an (001) Si wafer grown by MBE at 650 °C. Images (a) and (b) are pseudo-3D and grey-scale views of a sample that exhibited both pyramidal and dome-shaped island formation. Image (c) is from a sample that exhibited large faceted domes. The corresponding contour lines in the directions of the arrows are shown in panels (d), (e), and (f). (Adapted from Baribeau et al., 2005, *J. Mater. Res.*, with permission from the Materials Research Society.)

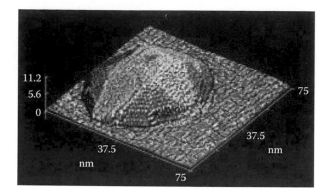

FIGURE 8.7 An atomic-resolution STM topographic image of a faceted Ge dome grown by PVD on a Si (001) substrate held at 600 °C. (Adapted from Williams et al., 1999, *Acc. Chem. Res.*, with permission from the American Chemical Society.)

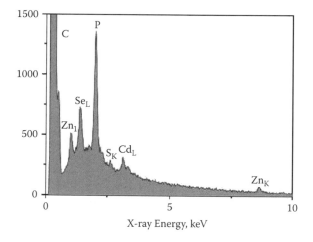

FIGURE 8.8 EDS data obtained from ZnSe/CdS core-shell dots where the shell consisted of four monolayers of CdS. The presence of phosphorus is adventitious. (From Nemchinov et al., 2008, *J. Phys. Chem. C*, with permission from the American Chemical Society.)

8.4.2 X-Ray Diffraction and Scattering Methods for Ensembles—XRD, SAXS, and WAXS

Characterization by routine XRD of the crystal structure of quantum dot powder specimens can be a useful screening tool. On the other hand, small-angle X-ray scattering (SAXS) and wide-angle X-ray scattering (WAXS) cannot be considered to be routine techniques, being based either on high-power rotating anode instruments, or on synchrotron sources.

8.4.2.1 Characterization by XRD

Figure 8.9 shows results obtained by XRD for the bulk CdSe wurtzite structure, as compared with powder specimens with size fractions of 2.8, 4.1, and 5.6 nm. The

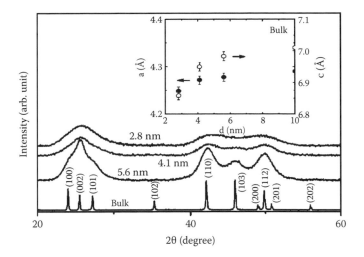

FIGURE 8.9 X-ray diffraction data for bulk wurtzite CdSe (lower trace) and for powder specimens with particle diameters of 2.8, 4.1, and 5.6 nm, respectively. The inset shows changes in *a*- and *c*-axis d-spacings as a function of particle size. (From Neeleshwar et al., 2005, *Phys. Rev. B*, with permission from the American Physical Society.)

expected line broadening with decreasing particle diameter is evident. As well, a decrease in d-spacings is observed (see inset) with decreasing diameter (Neeleshwar et al. 2005).

XRD line broadening due to crystallite size, or particle size, *d*, is described by the Scherrer equation,

$$d = \frac{K\lambda}{B\cos\theta_B} \tag{8.26}$$

where *K* is the Scherrer constant (ca. unity), *B* is the half-width of the Bragg peak, λ is the wavelength of the radiation, and θ_B is the Bragg angle. An example is shown in Figure 8.10 for a size range of CdSe particles.

8.4.2.2 Small-Angle X-Ray Scattering (SAXS) of Quantum Dots

Average particle size and size distribution down to a few nm can be determined by a number of well-established and convenient techniques (see Chapter 6). The main short-coming of these techniques is that their analyses are based on the assumption of an 'equivalent sphere'.

The SAXS technique has been used principally to investigate the structures of macromolecules and clusters, but it can also usefully be applied to particles in the size range up to ca. 25 nm. In general, the technique requires a high-luminosity collimated X-ray source, and is therefore most commonly attached to a synchrotron facility.

The value of the technique for the analysis of quantum dots can best be illustrated by an example from the literature. The discussion will follow the work of Murray et al. 2000.

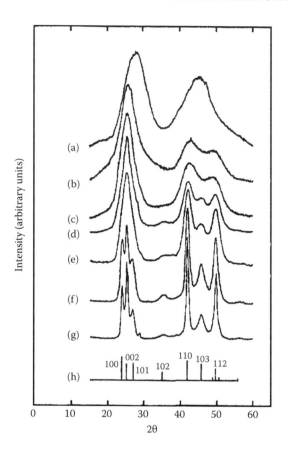

FIGURE 8.10 XRD line broadening illustrated by the powder patterns obtained from CdSe particles of diameters (a) 1.2, (b) 1.8, (c) 2.0, (d) 3.7, (e) 4.2, (f) 8.3, and (g) 11.5 nm compared with (h) the bulk wurtzite peak positions. (From Murray et al., 1993, *J. Am. Chem. Soc.*, with permission from the American Chemical Society.)

The SAXS intensity $I(q)$, scattered from a collection of N non-interacting particles, of uniform electron density ρ, and radius R, embedded in a homogeneous medium of density ρ_0, is given by

$$I(q) = I_0 N(\rho - \rho_0)^2 F^2(q) \tag{8.27}$$

where the form factor, $F(q)$, for a sphere is given by

$$F(q) = \frac{4\pi R^3}{3} \left[3\frac{\sin(qR) - qR\cos(qR)}{(qR)^3} \right] \tag{8.28}$$

The scattering parameter, q, is defined by

$$q = \frac{4\pi \sin \theta}{2\lambda} \qquad (8.29)$$

where θ is the scattering angle, and λ is the wavelength of the incident radiation.

The form factor is the Fourier transform of the shape of the scattering object (Glatter and Kratky 1982). It was used to calculate patterns for a collection of hypothetical 6.2 nm monodisperse spheres of uniform electron density. The real objects are unlikely to have uniform electron density. Each atom in the structure will contribute a modulation that adds a broad scattering background. Curves in Figure 8.11 show the scattering patterns for a spherical fragment containing 4500 atoms with a bulk CdSe wurtzite structure. It corresponds to the sum of the scattering from atomic sites and the coherent scattering from all discrete spacings within the particle.

Scattering from atomic sites scales with R^{-3}, while the intensity of the oscillations from the particle shape scales as R^{-6}. In the context of the SAXS analysis of particles in the size range of a few nm, the atomic site scattering is generally non-negligible, and must be considered in order avoid an overestimation of the size distribution.

TEM studies have shown that many nanoparticles are better described as ellipsoids, rather than as spheres (e.g., Murray, Norris, and Bawendi 1993). Accordingly, a refined atomistic model, based on ellipsoidal fragments, was used to 'correct' the SAXS scattering. The amplitude of the oscillations allows the determination of particle size and size distribution. The modelling showed the predicted pattern for a monodisperse ensemble of prolate (with an aspect ratio of 1.2 as indicated by TEM analysis) fragments of the bulk CdSe lattice. These results were fitted to the experimental data for a best fit to an atomistic model of SAXS scattering, based on the observations by TEM. Experimental SAXS patterns (dots) and computer simulations (solid lines) were then used to measure size and size distribution for a series of CdSe particles (Figure 8.11). A companion study based on the use of a related WAXS technique has also been described by Murray, Kagan, and Bawendi (2000).

8.5 ABSORPTION AND PHOTOLUMINESCENCE SPECTROSCOPY OF QUANTUM DOTS

These two spectroscopies, in which the photoluminescence process is the combination of excitation by incident photons and the subsequent de-excitation via a fluorescence decay channel, are arguably the most convenient and widely used tools for screening products from the various synthesis routes for quantum dots. The combination of the two spectroscopies addresses most of the issues that are related to the electronic structure.

From the absorption spectrum, over the UV-VIS spectral range, the exciton formation energy, required when an electron in the ground state is promoted to an excited state, can be obtained; while from the fluorescence spectrum, generally over the VIS-IR range, one or more of the decay channels for which the relaxation, (leading to recombination), is accompanied by emission of a photon with characteristic energy, can be deduced.

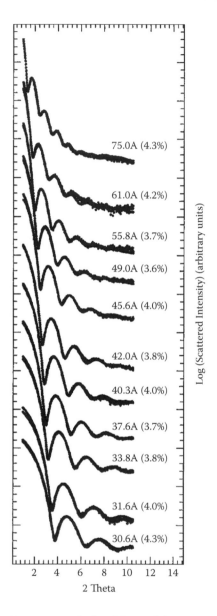

75.0A (4.3%)

61.0A (4.2%)

55.8A (3.7%)

49.0A (3.6%)

45.6A (4.0%)

42.0A (3.8%)

40.3A (4.0%)

37.6A (3.7%)

33.8A (3.8%)

31.6A (4.0%)

30.6A (4.3%)

Log (Scattered Intensity) (arbitrary units)

2   4   6   8   10   12   14

2 Theta

FIGURE 8.11 SAXS patterns were calculated for model structures containing 4500 atoms, comparable to a CdSe spherical particle of 6.2 nm diameter (not shown). The curves represented models for 6.2 nm spheres of uniform electron density; monodisperse spherical fragments of the bulk CdSe structure; monodisperse ellipsoidal fragments of the bulk CdSe structure, having an aspect ratio of 1.2; and a best fit to SAXS data assuming a Gaussian distribution of ellipsoids, yielding the particle size and size distribution. The curves in the figure show SAXS patterns for CdSe particle samples ranging from 3 to 75 nm in diameter (dots). The best fits are used to infer the particle size for each batch, in terms of equivalent diameters, and size distributions (the width of the distribution divided by the mean diameter, in %), ranging from 3.5% to 4.5% for the samples shown. (Adapted from Murray et al., 2000.)

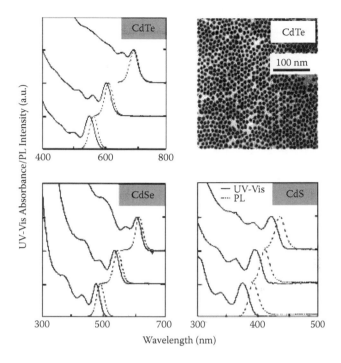

FIGURE 8.12 Absorption and photoluminescence spectroscopic data from three quantum dot compositions, each with three particle diameters in the size range 2–5 nm. The vertical axis is in arbitrary units for the UV-Vis absorption data (solid curves), and for photoluminescence emission (broken curves). The expected shifts to lower wavelength, that is, greater energy gap between ground state and excited state, with decreasing particle diameters can be seen (From Yu et al., 2003, *Chem. Mater.*, with permission from the American Chemical Society.)

The combined spectroscopies can determine the 'bandgap' for a compound semiconductor dot, or the energy of the excited states with respect to the ground state. As well, the effects of electrically active impurities and defects can be revealed, with their respective concentrations being related to the intensities of the spectral structures. Time-resolved luminescence spectroscopy can shed light on recombination processes, and provide insight into the 'blinking' phenomenon. As a screening tool, fluorescence spectroscopy can discriminate between 'good' and 'bad' outcomes from any particular synthesis route. Some illustrative examples will be considered below.

Absorption and photoluminescence (i.e., fluorescence) spectra from CdTe, CdSe, and CdS quantum dots, as functions of particle size, are shown in Figure 8.12. A TEM image of the CdTe dots is also shown. The sizes for the three batches of each composition were in the range 2–5 nm. The shift to shorter wavelength (higher energy) with decreasing particle size is apparent, in accord with the dependence of level spacing on dimension of region of confinement (see discussion of quantum wells above) (Yu et al. 2003).

Similar results for the $CdSe_{0.34}Te_{0.66}$ composition, for a wider size range, are shown in Figure 8.13. The CdSe/ZnS core-shell quantum dot system has been characterized with absorption/luminescence spectroscopy by Dabbousi et al. (1997).

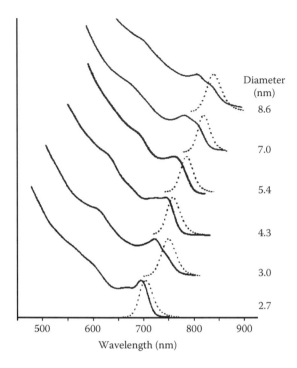

FIGURE 8.13 Absorption (solid lines) and photoluminescence (broken lines) spectra as functions of diameter of $CdSe_{0.34}Te_{0.66}$ QDs. (Adapted from Bailey and Nie, 2003, *J. Am. Chem. Soc.*, with permission from the American Chemical Society.)

It has been found (e.g., Nirmal et al. 1996) that under continuous illumination the photoluminescence of the majority of single quantum dots undergoes intermittent changes of intensity from 'on' to 'off'. The phenomenon is commonly referred to as 'blinking', see Figure 8.14. The issue has attracted some interest, due to the availability of time-resolved information obtainable from single quantum dots. Recent experimental results and theoretical approaches to descriptions of blinking can be found in the literature (e.g., Cichos, Martin, and von Borczyskowski 2004; Peterson and Krauss 2006; Gómez, Califano, and Mulvaney 2006).

8.5.1 PHOTOLUMINESCENCE OF SINGLE QUANTUM DOTS

Ensembles of nominally monodisperse quantum dots do, in fact, have a finite, although possibly small, spread in diameter. Thus, there will be a corresponding spread in the individual energy levels of the ensemble of dots. As a result, photoluminescence spectra will exhibit broad averaged peaks within which the fine structure has been degraded. The effect is illustrated in Figure 8.15. Therein lies an important motivation for the ability to interrogate the spectra of single quantum dots. Development of confocal optical spectroscopy was the first approach towards obtaining spectra from single quantum dots. An example is shown in Figure 8.16 of spectroscopic data obtained from a single dot.

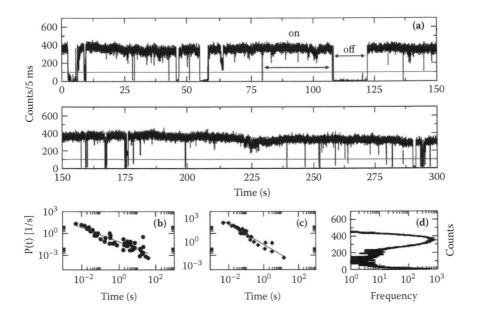

FIGURE 8.14 (a) Time-resolved photoluminescence intensity for a single CdSe/ZnS core-shell dot continuously excited by a 488 nm laser with an incident power of ca. 0.6 kW cm⁻². The trace of duration 300s illustrates irregular blinking events, seemingly of random dura-tion.. The light grey line at 100 counts/ms discriminates between on and off events. (b and c) However, the log–log plots of frequency vs. duration of on- and off-times, respectively, exhibit linear dependencies (i.e., both on- and off-events of short duration are favoured in comparison with events of long duration. (d) The histogram of the intensity of the trace in (a) suggests that there is maximum probability at zero intensity (presumably corresponding to the system being in the ground state) and for a higher intensity of ca. 400 counts (presumably corresponding to the system then undergoing decay from the first excited state). (Adapted from Gómez et al., 2006, *Phys. Chem. Chem. Phys.*, with permission from the American Physical Society.)

A more recent development for single dot spectroscopy is based on scanning near-field optical microscopy (SNOM) (see Chapter 5 for technical details). The spatial decay of near-field radiation from a source defined by a small aperture (typically 50 nm in diameter) can provide spatial resolution comparable to the diameter of the aperture. The images in Figure 8.17 demonstrate the available resolution as a func-tion of aperture size.

Spatial mapping of individual decay modes, as well as single quantum dot spec-troscopy, by near-field spectroscopy is shown in Figure 8.18.

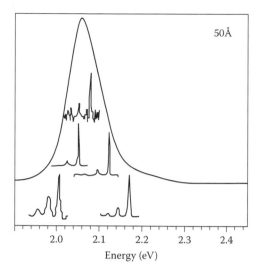

FIGURE 8.15 Schematic illustration of an ensemble spectrum, which, in fact, is a convolution of several single quantum dot spectra shifted in energy due to the finite spread in size of a nominally monodisperse powder sample. (From Empedocles et al., 1999, with permission from John Wiley and Sons.)

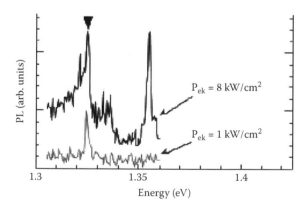

FIGURE 8.16 High spectral resolution photoluminescence spectra from a single quantum dot obtained at low temperature for two levels of excitation power. (Adapted from Roussignol S., Low Temperature Photoluminescence (PL) Spectroscopy on Quantum Dots, http://www.oxinst.com/products/low-temperature/applications-library/optical-spectroscopic/ Documents/MicrostatHiResII-Low-temperature-photoluminescence-spectroscopy-on-quantum-dots-application-note.pdf.)

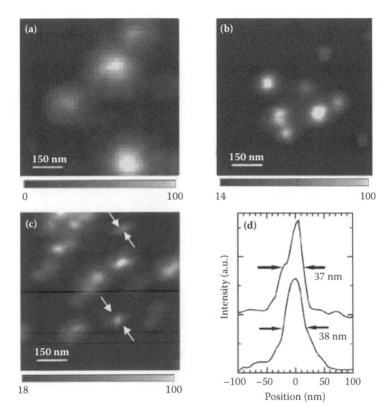

FIGURE 8.17   Near-field optical images of single InAs quantum dots. Images (a–c) were recorded for apertures of 135, 75, and 30 nm, respectively. The intensity traces taken from (c) suggest a resolution of ca. 30 nm, corresponding to the dot diameter. The dependence of spatial resolution on size of aperture is apparent. (From Matsuda et al., 2002, *Appl. Phys. Lett.*, with permission from the American Physical Society.)

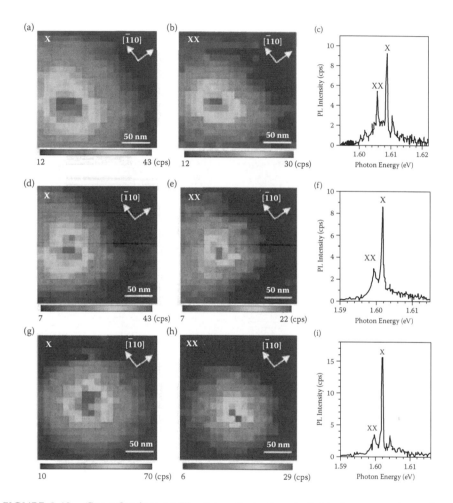

FIGURE 8.18 (**See color insert.**) The data refer to three individual quantum dots, where the exciton decay proceeds through two channels, giving rise to emission lines labelled X and XX. The intensity of emission line X is mapped for the three dots in a, d, and g, while the XX line is mapped in b, e, and h. The integrated intensities for the three dots are shown in spectral form in c, f, i. (From Matsuda et al., 2003, *Phys. Rev. Lett.*, with permission from the American Physical Society.)

REFERENCES

Bailey, R. E. and Nie, S., 2003, *J. Am. Chem. Soc.*, 125, 7100.

Baribeau, J. M., Rowell, N. L., and Lockwood, D. J., 2005, *J. Mater. Res.*, 20, 3278.

Brunner, K., 2002, *Rep. Prog. Phys.*, 65, 27.

Cichos, F. Martin, J. and von Borczyskowski, C., 2004, *Phys. Rev. B*, 70, 115314.

Dabbousi, B. O., Rodriguez-Viejo, J., Mikulec, F. V, Heine, J. R., Mattoussi, H., Ober, R., Jensen, K. F., and Bawendi, M. G., 1997, *J. Phys. Chem.*, 101, 9463.

Delerue, C. and Lannoo, M., 2004, *Nanostructures: Theory and Modelling,* Berlin: Springer-Verlag. Eaglesham D. J. and Cerullo, M. 1990, *Phys. Rev. Lett.,* 64, 1943.

Eaglesham, D. and Cerullo, M., 1990, *Phys. Rev. Lett.,* 64, 1943.

Eisberg, R. M. and Resnick, R., 1974, *Quantum Physics of Atoms, Molecules, Solids, Nuclei and Particles,* New York, NY: Wiley.

Empedocles, S. A., Neuhauser, R., Shimizu, K., and Bawendi, M. G., 1999, *Adv. Mater.,* 11, 1243.

Glatter, O. and Kratky, O., 1982, *Small Angle X-Ray Scattering*, New York: Academic Press.

Gómez, D. E., Califano, M., and Mulvaney, P., 2006, *Phys. Chem. Chem. Phys.*, 8, 4989.

Griffiths, D. J., 2005, *Introduction to Quantum Mechanics*, 2nd ed., New York: Prentice Hall.

Jamieson, T., Bakhshi, R., Petrova, D., Pocock, R., Imani, M., and Seifaliana, A. M., 2007, *Biomaterials*, 28, 4717.

Kouwenhoven, L. P., Austing, D. G., and Tarucha, S., 2001, *Rep. Prog. Phys.*, 64, 701.

Lindberg, V. and Hellsing, B., 2005, *J. Phys. Condens. Matter*, 17, S1075.

Manna, L., Scher, E. C., and Alivisatos, A. P., 2002, *J. Cluster Sci.*, 13, 521.

Matsuda, K., Saiki, T., Nomura, S., Mihara, M., and Aoyagi, Y., 2002, *Appl. Phys. Lett.*, 81, 2291.

Matsuda, K., Saiki, T., Nomura, S., Mihara, M., Aoyagi, Y., Nair, S., and Takagahara, T., 2003, *Phys. Rev. Lett.*, 91, 177401.

Michalet, X., Pinaud, F. F., Bentolila, L. A., Tsay, J. M., Doose, S., Li, J. J., Sundaresan, G., Wu, A. M., Gambhir, S. S., and Weiss, S., 2005, *Science*, 307, 538.

Mo, Y.W., Savage, D. E., Swartzentruber, B. S., and Lagally, M. G., 1990, *Phys. Rev. Lett.,* 65, 1020.

Murray, C. B., Kagan, C. R., and Bawendi, M. G., 2000, *Ann. Rev. Mater. Sci.*, 30, 545.

Murray, C. B., Norris, D. J., and Bawendi, M. G., 1993, *J. Am. Chem. Soc.*, 115, 8706.

Neeleshwar, S., Chen, C. L., Tsai, C. B., Chen, Y. Y., Chen, C. C., Shyu, S. G., and Seehra, M. S., 2005, *Phys. Rev. B*, 71, 201307.

Nemchinov, A., Kirsanova, A., Hewa-Kasakarage, N. N., and Zamkov, M., 2008, *J. Phys. Chem. C*, 112, 9301.

Nirmal, M., Daboussi, B. O., Bawendi, M. G., Macklin, J. J., Trautman, J. K., Harris, T. D., and Brus, L. E., 1996, *Nature*, 383, 802.

Peterson, J. J. and Krauss, T. D., 2006, *Nano Lett.,* 6, 510.

Reimann, S. M. and Manninen, M., 2002, *Rev. Mod. Phys.,* 74, 1283.

Rogach, A. L., Talapin, D. V., Shevchenko, E. V., Kornowski, A., Haase, M., and Weller, H., 2002, *Adv. Funct. Mater.*, 12, 653.

Roussignol S., Low Temperature Photoluminescence (PL) Spectroscopy on Quantum Dots, Paris, France: Pierre Aigrain Laboratory, Ecole Normale Supérieure, http://www.oxinst.com/products/low-temperature/applications-library/optical-spectroscopic/Documents/MicrostatHiResII-Low-temperature-photoluminescence-spectroscopy-on-quantum-dots-application-note.pdf.

Tersoff, J., Teichert, C., and Lagally, M. G., 1996, *Phys. Rev. Lett.*, 76, 1675.

Tsukamoto, S., Honma, T., Bell, G. R., Ishii, A., and Arakawa, Y., 2006, *Small*, 2, 386.

Vanmaekelbergh, D. and Liljeroth, P., 2005, *Chem. Soc. Rev.*, 34, 299.

Wang, L. W. and Zunger, A., 1996, *Phys. Rev. B*, 53, 9579.

Williams, R. S., Medeiros-Ribeiro, G., Kamins, T. I., and Ohlberg, D. A. A., 1999, *Acc. Chem. Res.*, 32, 425.
Yu, W. W., Qu, L., Guo, W., and Peng, X., 2003, *Chem. Mater.*, 15, 2854.

9 Carbon Nanotubes and Other Tube Structures

9.1 INTRODUCTION

Multi-wall CNTs were first observed and described in 1991 (Iijima 1991) during (high resolution) transmission electron microscopic (HRTEM) analysis of carbon arc electrodes, Figure 9.1. In recent years, other tube structures have been synthesised and described. However, the CNT variety has remained the focus of attention in the R&D community, and accordingly more prominence will be given to it in this chapter, in which methods for their characterization are described.

Subsequently, one-dimensional quantum effects were observed in accord with predictions based on their electronic properties. Although the initial experimental observations were for MWCNTs, single-wall carbon nanotubes (SWCNTs) had attracted theoretical interest and predictions that preceded their experimental observation. The most noteworthy predictions were that CNTs could be either semiconductors or metals depending on their characteristics, namely their diameters and the orientation of their hexagons with respect to the CNT axis (chiral angle) (Dresselhaus, Dresselhaus, and Saito 1992; Hamada et al. 1992; Mintmire, Dunlap, and White 1992). The predictions were corroborated experimentally several years later (Odom et al. 1998; Wildöer et al. 1998) (see Figure 9.2).

At the present time, CNTs are primarily research materials, and, like others, are produced in relatively small volumes at high cost. However, commercially viable products including tennis racquets with carbon nanotube reinforcement, and high-aspect-ratio probes for scanning probe microscopes (SPMs) are beginning to emerge.

9.2 DESCRIPTION OF CNT STRUCTURE

The CNT structure can be thought of as a cylindrical folding up of n graphene sheets where n = 1 corresponds to the single wall variety (SWCNT), and n > 1 to the multi-wall structures (MWCNT). More formally, an SWCNT can be viewed as the conformal mapping of an infinite strip of a graphene sheet onto a cylindrical surface. The chirality of a CNT can be characterized by a set of two integers (n_1, n_2, other forms of notation are used, such as (m,n)) corresponding to a lattice vector

$$\mathbf{L} = n_1\mathbf{a}_1 + n_2\mathbf{a}_2 \qquad (9.1)$$

where $\mathbf{a}_1$ and $\mathbf{a}_2$ are the primitive unit cell vectors in the basal plane of the graphite structure. The definition of the unit cell for CNTs refers to a 2-D space of a cylindrical section, rather than to a 3-D structure, in which a unit cell is defined by the

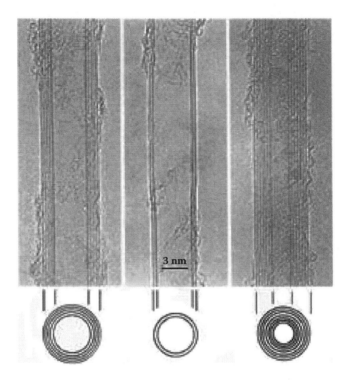

FIGURE 9.1 Original observations of multi-wall CNTs (MWCNTs) showing concentric nested structures. (From Iijima, 1991, *Nature*, reprinted by permission from Macmillan Publishers Ltd.)

requirement that it must be capable of filling all space by repetitive translation by unit cell vectors. The chirality of the structure is defined by a vector $OA = C_h$ perpendicular to the long axis OB of the tube, see Figure 9.3. There are two particular chiralities familiar to chemists:

For $n_2 = 0$ the result is a zig-zag structure,
For $n_1 = n_2$ the result is an arm-chair structure,
If $n_1 \neq n_2$, and $n_2 \neq 0$, then the tube is said to be 'chiral'.

The chiral angle θ is the angle between L and $\mathbf{a}_1$. The basic symmetry operation $S = (\Psi/\tau)$ is an angle Ψ of rotation around the long axis followed by a displacement τ along the axis.

The unit cell length along the axis of the tube is given by T, where T is perpendicular to L.

$$T = \frac{(n_1 + 2n_2)a_1 - (2n_1 + n_2)a_2}{q} \tag{9.2}$$

where $q = N$ if $(n_1 - n_2) \neq \mod(3)$, or $= 3N$ if $(n_1 - n_2) = \mod(3)$, and where N is the greatest common divisor of n_1 and n_2. The tube radius is given by

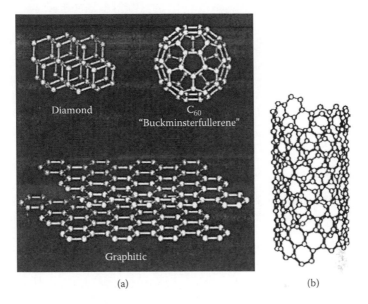

FIGURE 9.2 The inventory of principal carbon-based phases includes (a) diamond, graphite and C_{60}, as well as (b) carbon nanotube structures. In each case, there are variations on the theme.

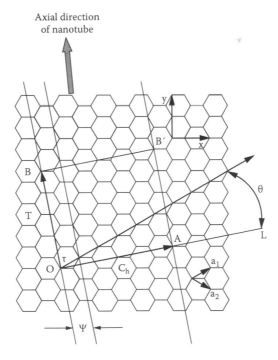

FIGURE 9.3 Schematic of descriptive structural parameters for CNT. The diagram represents the $(n_1, n_2) = (4,2)$ configuration.

$$R = \frac{L}{2\pi} = \frac{\sqrt{3}d}{2\pi}(n_1^2 + n_2^2 + n_1 n_2)^{1/2} \tag{9.3}$$

where $d = 0.142$ nm (the C-C bond length). In general, CNTs are produced as multi-wall co-axially nested structures with a layer spacing of 0.34 nm, similar to that of the interlayer spacing in graphite.

9.3   SYNTHESIS ROUTES

9.3.1   ARC-DISCHARGE AND LASER ABLATION

The arc-discharge and laser ablation methods for the growth of CNTs are based on the condensation of carbon atoms generated from evaporation of solid carbon sources. The temperature of evaporation is in the range 3000–4000°C, which is close to the melting temperature of graphite.

In an arc-discharge, carbon atoms are evaporated by a helium plasma initiated by high currents passing between the opposing carbon electrodes, anode and cathode. Arc-discharge has been developed into a reasonably reliable method for producing both MWCNTs and SWCNTs, but for the growth of SWCNTs, a metal catalyst is needed in the arc-discharge system. The first success in producing substantial quantities of SWCNTs by arc-discharge was achieved by Bethune and coworkers (1993), who used a carbon anode containing a small percentage of cobalt.

The growth of high quality, in terms of yield, purity, defects, length, and linearity, SWCNTs, was achieved by Thess et al. in 1996 using a laser ablation method, in which intense laser pulses were directed at a carbon target containing 0.5 at.% of nickel and cobalt, as catalysts. The target was placed in a tube-furnace heated to 1200°C. The arc-discharge process in the presence of a catalyst appears to favour the growth of SWCNTs with amorphous carbon as the predominant impurity. The highest yield reported so far has involved the use of a mixture of 2.8:1 of Ni:Y as the catalyst, mixed with graphite powder and placed inside a hollow graphite electrode (Farhat 2001).

In the case of SWCNT growth by arc-discharge and laser ablation, typical by-products include fullerene, graphitic polyhedrons with enclosed metal particles, and amorphous carbon in the form of particles or as a coating on the CNT sidewalls.

9.3.2   CHEMICAL VAPOUR DEPOSITION (CVD)

Growth of CNTs by CVD involves a catalyst material exposed to a flowing hydrocarbon gas at high temperatures in a furnace. Materials that have grown on the catalyst are collected after cooling the system to room temperature. The main operating parameters include the partial pressure of hydrocarbons, the presence or absence of a carrier gas, the type of catalyst, substrate temperature and temperature gradients, and the length of exposure time. The active catalytic species are typically in the form of transition-metal nanoparticles carried on a support material such as alumina, and the general CNT growth mechanism involves the dissolution and saturation of carbon atoms in the metal nanoparticles. The precipitation of carbon from the saturated metal particle leads to the formation of tubular carbon solids with sp^2 bonding. Tubular formation is favoured over other forms of carbon such as graphitic sheets with open edges, because a tube contains no dangling bonds and, therefore, is a preferred low-energy configuration.

Recent interest in the CVD growth of CNTs arises from the idea that aligned and ordered CNT structures can be grown on surfaces in a controlled manner not possible with arc-discharge or laser ablation techniques (Dai et al. 1999).

Methods developed to produce MWCNTs have included CVD growth of tubes in the pores of mesoporous silica. Dai et al. (1999) have devised growth strategies for ordered MWCNTs and SWCNTs by CVD on a catalytically patterned substrate. They found that MWCNTs can self-assemble into aligned structures as they grow, with the driving force for self-alignment being the van der Waals interaction between them. This growth approach involves catalyst patterning and careful preparation of the substrate to enhance catalyst-substrate interaction and control of the catalyst particle size.

Ordered SWCNT structures have been grown by the CVD method with methane as the feedstock on catalytically patterned substrates (e.g., Cassell et al. 1999; Franklin and Dai 2000).

9.3.2 Purification of Raw Product from Synthesis Routes

The raw material present directly after synthesis (prior to any purification steps) usually contains amorphous carbon, graphitic particles, metal catalytic particles, and often only a very small number of CNTs. In general, both SWCNTs and MWCNTs will appear in the same sample batch. Accordingly, purification is normally performed in order to improve the quality of the final product, and characterization has a vital part to play in ensuring quality control.

Purification of the raw material requires the removal of the impurities (generally arising from either or both the substrate material and catalyst particles) and unwanted carbonaceous phases. The removal of the metal catalyst and alumina substrate typically involves a chemical process consisting of treatment with acids (HCl and HNO_3), and often with heating or microwave exposure as well. This is normally followed by washing with solvents and filtration. The removal of the graphitic particles and layers of amorphous carbon adhering to the CNT's walls involves a more lengthy procedure of steam treatment, which can also remove the end cap of the CNT (e.g., Tobias et al. 2006; Shao et al. 2007). Removal of amorphous carbon is an essential requirement for functionalising CNT structures. Functional groups such as hydroxyl groups, carboxylic groups, quinones, phenols, and/or lactones may be attached to the SWNT sidewalls and/or at the ends of the tubes during the purification by oxidation with strong acids. Of course, precise understanding of the nature of these functionalities, in addition to the estimation of functionalisation level and overall sample purity and homogeneity is very desirable.

9.4 ELECTRONIC STRUCTURE OF GRAPHENE AND SWCNT

The energy dispersion relations of SWCNTs can be calculated using zone folding (Hamada et al. 1992; Jishi et al. 1994), the tight-binding method (Popov 2004), density functional theory (DFT) (Reich and Ordejón 2002; Zólyomi and Kürti 2004), and linear combination of atomic orbitals (LCAO). Qualitative insight into the general features of the electronic structure of graphene and SWCNT can be gained by the simple method of zone folding the structure of graphene obtained by the tight-binding method.

9.4.1 ELECTRONIC BAND STRUCTURE OF GRAPHENE

The structure of graphene is shown in Figure 9.4.

The distance between points 1 and 2 and the components of the primitive lattice vectors are given by

$$|a_{1,2}| = \sqrt{3}a_0$$

$$a_1 = \sqrt{3}a_0 \left(\frac{1}{2}, \frac{\sqrt{3}}{2} \right)$$

$$a_2 = \sqrt{3}a_0 \left(-\frac{1}{2}, \frac{\sqrt{3}}{2} \right)$$

(9.4)

Within the tight-binding method, the two-dimensional energy dispersion relations of graphene can be calculated by solving the eigenvalue problem for a Hamiltonian, $H_{g\text{-}2D}$, associated with the two carbon atoms, 1 and 2, in the graphene unit cell. In the Slater–Koster scheme, the Hamiltonian can be written

$$H_{g\text{-}2D} = \begin{vmatrix} 0 & f(k) \\ -f*(k) & 0 \end{vmatrix}$$

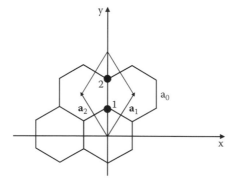

FIGURE 9.4 Structure of graphene. Atoms are located at the intersections, and the sp² bonds connect the lattice points. The primitive unit cell is defined by vectors $\mathbf{a}_1$ and $\mathbf{a}_2$. There are two atoms per cell labelled 1 and 2.

where

$$f(k) = -E_{C-C}(1 + e^{ik \cdot a_1} + e^{ik \cdot a_2}) = -E_{C-C}(1 + 2e^{\sqrt{3}k_x a/2} \cos(k_y a/2)) \qquad (9.5)$$

and k_x and k_y are wave vectors in the x- and y-directions.

Only nearest neighbour overlap energies, E_{C-C}, are included. The solution of the secular equation

$$\text{Det } |H_{g\text{-}2D} - EI| \qquad (9.6)$$

gives

$$E_{g\text{-}2D}^{\pm}(k) = \pm E_{C-C}\left[1 + 4\cos\left(\frac{\sqrt{3}k_x a}{2}\right)\cos\left(\frac{k_y a}{2}\right) + \cos^2\left(\frac{k_y a}{2}\right)\right]^{1/2} \qquad (9.7)$$

where $E_{g\text{-}2D}^{+}(k)$ and $E_{g\text{-}2D}^{-}(k)$ correspond to the π^* and π energy bands, respectively.

9.4.2 ELECTRONIC STRUCTURE OF SWCNTs

The electronic structure of an SWCNT can be obtained from that of graphene. Periodic boundary conditions can be applied for the circumference characterized by the chiral vector, C_h. Accordingly, the wave vector associated with the direction of C_h becomes quantised, while that associated with the direction of the translational vector remains continuous for a CNT of infinite length. Thus, the energy bands consist of a set of one-dimensional energy dispersion relations, which are cross-sections of those of graphene. The expressions for the reciprocal lattice vectors G_2 along the CNT axis, $(C_h \cdot G_2 = 0,$ and $T \cdot G_2 = 2\pi)$, and G_1 are

$$G_1 = \frac{-t_2 b_1 + t_1 b_2}{N}$$

$$\qquad (9.8)$$

$$G_2 = \frac{n_2 b_1 - n_1 b_2}{N}$$

where N is the number of hexagons in the unit cell. The G-vectors are constructed from the primitive reciprocal vectors b_1 and b_2, defined in Table 9.1.

The one-dimensional energy dispersion relations of an SWCNT can be written as

$$E_{CNT}^{v}(k) = E_{g\text{-}2D}\left(k\frac{G_2}{|G_2|} + vG_1\right) \qquad (9.9)$$

where $-\pi/T < k < \pi/T$ refers to the one-dimensional wave vector along the CNT axis and $v = 1, 2, \ldots, N$. The periodic boundary condition for a CNT gives N discrete k

TABLE 9.1
Summary of Structural Parameters for CNTs

Symbol	Description	Formula
a	Unit vector length	$a = \sqrt{3}a_{C-C} = 0.249$ nm $\\ a_{C-C} = 0.142$
$\boldsymbol{a}_1, \boldsymbol{a}_2$	Unit vectors	$\left(\dfrac{\sqrt{3}}{2}, \dfrac{1}{2}\right)a, \left(\dfrac{\sqrt{3}}{2}, -\dfrac{1}{2}\right)a$
$\boldsymbol{b}_1, \boldsymbol{b}_2$	Reciprocal lattice vectors	$\left(\dfrac{1}{\sqrt{3}}, 1\right)\dfrac{2\pi}{a}, \left(\dfrac{1}{\sqrt{3}}, -1\right)\dfrac{2\pi}{a}$
$\boldsymbol{C}_h$	Chiral vector	$\boldsymbol{C}_h = n_1\boldsymbol{a}_1 + n_2\boldsymbol{a}_2 \equiv (n_1, n_2) \\ (0 \le |n_2| \le n_1)$
L	Length of C_h	$L = |\boldsymbol{C}_h| = a(n_1^2 + n_2^2 + n_1 n_2)$
d_{CNT}	Diameter of tube	$D_{CNT} = L/\pi$
θ	Chiral angle	$\tan(\theta) = \dfrac{\sqrt{3}}{2n_1 + n_2}$
d_R	—	Greatest common denominator of $(2n_1 + n_2, 2n_2 + n_1)$
T	Translational vector	$\boldsymbol{T} = t_1\boldsymbol{a}_1 + t_2\boldsymbol{a}_2 \equiv (t_1, t_2) \\ t_1 = \dfrac{2n_2 + n_1}{d_R},\ t_2 = -\dfrac{2n_1 + n_2}{d_R}$
T	Length of T	$T = |\boldsymbol{T}| = \dfrac{\sqrt{3}L}{d_R}$
N	Number of hexagons in unit cell	$N = \dfrac{3(n_1^2 + n_2^2 + n_1 n_2)}{d_R}$

Source: Saito et al., 1998, *Physical Properties of Carbon Nanotubes*, London: Imperial College Press.

values in the direction of $\boldsymbol{C}_h$. The N pairs of energy dispersion curves from Equation (9.9) correspond to the cross-sections of the two-dimensional energy dispersion surface of graphene. In Figure 9.5, several cutting lines near one of the $\boldsymbol{K}$-points are shown. The separation between two adjacent lines and the lengths of the cutting lines are given by $|\boldsymbol{G}_1| = 2/d_{CNT}$ and $|\boldsymbol{G}_2| = 2\pi/T$, respectively. If the cutting line passes through a $\boldsymbol{K}$ point of the two-dimensional Brillouin zone, where the π and π^* energy bands of graphene are degenerate by symmetry, then the one-dimensional energy bands have a zero energy gap. When the $\boldsymbol{K}$-point is located between two cutting lines, $\boldsymbol{K}$ is always located in a position one-third of the distance between two adjacent $\boldsymbol{G}_2$ lines (Jishi et al. 1994), and thus, a semiconducting CNT with a finite energy gap is formed. If for a (n_1, n_2) CNT, $(n_1 - n_2)$ is exactly divisible by 3, then the CNT is metallic. CNTs with residuals 1 and 2 from the division of $(n_1 - n_2)$ by 3 are semiconducting.

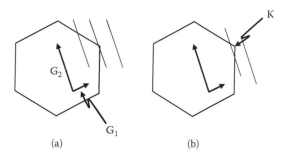

FIGURE 9.5 The one-dimensional *k*-vectors are shown overlaid on the Brillouin zone of graphene. The two plots illustrate the cases of (a) a metallic CNT, when the *G*-vector intersects the *K*-point of the zone, and (b) a semiconducting CNT, when the *G*-vector cuts the zone boundary.

In general (n,n) armchair CNTs have $4n$ energy sub-bands with $2n$ conduction and $2n$ valence bands. Of these $2n$ bands, two are non-degenerate and $n-1$ are doubly degenerate. The degeneracy comes from the two sub-bands with the same energy dispersion, but different v-values. All armchair CNTs exhibit a band degeneracy between the highest valence and the lowest conduction band. In the case of zigzag CNTs, the lowest conduction and the highest valence bands are doubly degenerate.

In armchair and zigzag CNTs, the bands are symmetric with respect to $k = 0$. Since the band of an armchair CNT has a minimum at point $k = 2\pi/3a$, it has a mirror minimum at point $k = -2\pi/3a$ and therefore two equivalent valleys are present around the point $\pm 2\pi/3a$. The bands of zigzag and chiral CNTs can have at most one valley. In armchair CNTs, the bands cross the Fermi level at $\pm 2\pi/3a$. Accordingly, they should exhibit metallic conduction (Saito, Dresselhaus, and Dresselhaus 1998).

Electrical conduction is determined by states around the Fermi energy. The band structures and densities of states are shown in Figure 9.6 at the locations of minima for metallic and semiconducting SWCNTs.

9.5 GENERAL CHARACTERISTICS OF CNTS

In Table 9.2 are summarised some of the salient characteristics of CNT structures in comparison with those of some common engineering materials. The comparison serves to illustrate the exceptional combination of properties of tube structures.

9.6 OTHER TUBE STRUCTURES

A boron nitride nanotube (BNNT) is a structural analogue of a CNT; its synthesis and identification as such was achieved in 1995 (Chopra et al. 1995), see Figure 9.7. The pure and stoichiometric BNNT structure is an electrical insulator with a bandgap of ca. 5.5 eV, which, unlike the CNT, is essentially independent of tube chirality and morphology (Blasé et al. 1994). In addition, it is, in accord with the layered BN structure, much more thermally and chemically stable than a graphitic CNT structure (Pouch and Alterovitz 1990).

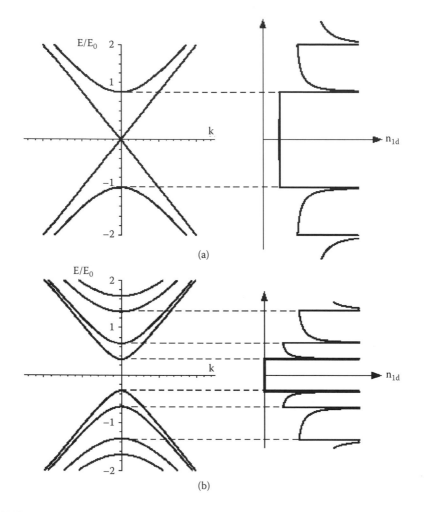

FIGURE 9.6   Schematic 1-D band structure (left) of an SWCNT and density of states (right). The situation for a metallic SWCNT is shown in (a), while that of a semiconducting one is shown in (b). In the latter case, the band gap is proportional to the diameter of the tube.

TABLE 9.2
Comparison of Characteristics of CNTs and of Other Materials

	SWCNT	Others
Strength (Young's modulus)	ca. 1 TPa	Stainless Steel: 0.18–0.24 TPa
Tensile Strength	13–53 GPa	Stainless Steel: 0.38–1.55 GPa
Hardness	462–546 GPa	Diamond: 150 GPa
Electrical Current	Metallic CNT: ca. 4×10^9 A/cm^2	Copper: ca. 4×10^6 A/cm^2
Thermal Conductivity	6000 W/mK	Copper: 385 W/mK

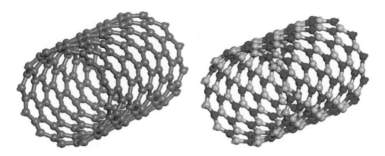

FIGURE 9.7   (left) CNT structure. (right) The BNNT structure is a direct analogue of that of the CNT, with B (dark) and N (bright) occupying alternate C-atom locations in the hexagonal graphitic net (From Golberg et al., 2007, *Adv. Mater.*, reprinted with permission from John Wiley and Sons.)

The methods of synthesis that have been developed for CNTs are, in general, not suitable for reliable high-yield production of BNNTs. Instead, a great variety of alternative synthesis routes have been explored with various degrees of success (see review by Golberg et al. 2007).

Nanoporous TiO_2 has been known and investigated since the 1950s, and first-generation large-diameter tube structures based on TiO_2 were first reported in 2001 (e.g., Gong et al. 2001). Subsequent generations have been synthesised by various routes, leading to structures with walls consisting of a few monolayers of TiO_2 molecules, and with inner diameters of a few nm. Much of the interest in TiO_2 tubes is based on their catalytic and photocatalytic activity, and in this context it is relevant that the exposed specific surface area can be much greater, by a factor of 10 or more, than that of raw TiO_2.particles. The structure of TiO_2 tubes has been reported as mixed anatase-rutile.

More recently, other tube structures, based on WO_3 and Ti-Fe-O, WS_2 (Rothschild et al. 1999), MoS_2, and GaSe, have been reported. However, in the context of characterization, the requirements are less exacting than those needed for CNTs.

9.7   CHARACTERIZATION OF NANOTUBES

The following characteristics are essential in specifying carbon nanotubes for all areas of application.

- Distribution of lengths.
- Distribution of diameters.
- Orientation with respect to substrate.
- Wall thickness and diameter of open core space.
- Number of walls, that is, single-walled, double-walled, or multi-walled.
- Chemical purity, that is, presence or absence of catalyst particles.
- Batch purity, that is, presence or absence of other phases and of adventitious contamination.
- Symmetry, for example, chirality.
- Extent of agglomeration and dispersibility.

TABLE 9.3

Quality Control Procedures Adopted by NASA for CNT Batches

Attribute	Technique	Analysis
Purity	TGA*	Quantitative; residual mass after TGA in air 5°C/min to 800°C
	SEM/TEM	Qualitative; presence of amorphous carbon impurities
	EDS	Qualitative; metal content
	Raman	Qualitative; carbon impurities and damage/disorder
Thermal stability	TGA	Quantitative; burning temperature in TGA in air at 5°C/min to 800°C, d(mass)/dT max
Homogeneity	TGA	Quantitative; standard deviation at burning temperature and residual mass for 3–5 samples
	SEM/TEM	Qualitative; comparison of images
Dispersibility	Ultrasonic	Qualitative; time taken to disperse fully agitation (visual) low conc. SWCNT in DMF (for standard settings)
	UV/VIS/NIR	Quantitative; relative change in absorption spectra during ultrasonic treatment

Source: Adapted from Arepelli et al., 2004, Protocol for the Characterization of Single-Wall Carbon Nanotube Material Quality (NASA JSC protocol), *Carbon.*

* TGA = Thermo-gravimetric analysis.

- Surface chemistry, with or without deliberate functionalisation.
- Without, or with, end capping, and if with, type of end capping.

NASA has adopted a protocol for determining ensemble averages for CNT with relevance to technical applications in the aerospace industry (Arepelli et al. 2004) (see Table 9.3).

9.7.1 Topographical and Structural Characterization

When information on single, or a small number of, nanotubes is required, then electron-optical (SEM and TEM) and probe techniques (AFM and STM) are the obvious choices. In other cases, when average information from ensembles of nanotubes is more relevant, then XRD or neutron diffraction will provide structural information. Indirect structural information can also be obtained from vibrational spectroscopies, such as Raman and IR, where vibrational spectral features can be related to structural models.

9.7.1.1 CNT

The most convenient and cost-effective technique for topographical imaging on the nanoscale is that of FESEM, where the current generation of instruments has a point-to-point resolution of ca. 1 nm. For the purpose of imaging vertically aligned nanotubes on a substrate, it is always best to use FESEM, because of its higher resolution at lower accelerating voltages. Recent results for vertically aligned NTs nucleated at, and grown from, catalytically active particles dispersed on a substrate, illustrate the use of the technique (Park et al. 2008). The synthesis of oriented CNTs with known termination at the apex is particularly relevant to their applications as field emitters (for flat-screen technology) and in X-ray generation by microsources (e.g., Bonard et al. 2001; Kim et al. 2000; Park et al. 2007). A cross-sectional view of a sample can confirm vertical alignment, as seen in Figure 9.8. The image in that figure also reveals the lengths of the nanotubes, their linear uniformity, and their approximate diameters. In addition, the image shows the lifted catalyst particles terminating the tubes. A top view shows the average spacing between the CNTs to be ca. 500 nm (see Figure 9.9).

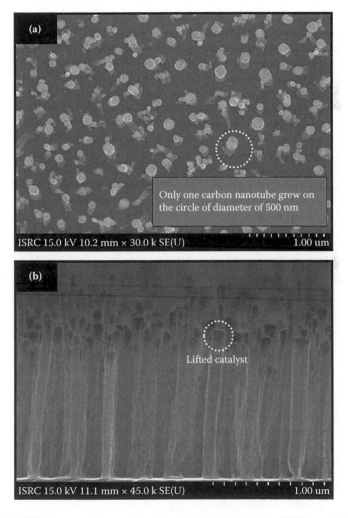

FIGURE 9.8 SEM images; (a) top view showing dispersed single CNTs, and (b) side view showing alignment, average lengths, and termination by catalyst particles. The lateral CNT dispersal was determined lithographically, and the objective was the production of a field emission source. (From Park et al., 2008, *Nanotechnology*, with permission from the Institute of Physics.)

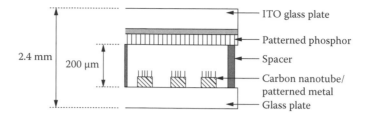

ITO glass plate

Patterned phosphor

Spacer

Carbon nanotube/
patterned metal

Glass plate

2.4 mm

200 μm

FIGURE 9.9 Conceptual schematic of a flat screen field emission display based on CNT emitters. (From Lee et al., 2001, *Diamond Relat. Materials*, with permission from Elsevier.)

FIGURE 9.10 SEM images of CNTs of various diameters grown on a nickel substrate. (From Milne et al., 2010, *J. Mater. Chem.*, reproduced with permission from the Royal Society of Chemistry.)

Vertical alignment of CNTs is a necessary, but not sufficient, requirement for use in field-emission displays. When the array of field-emitters on a CNT screen becomes excessively close-packed, there is a reduction in enhanced field emission. Likewise the effective radius of curvature at the apex is an important parameter. The images in Figures 9.10 and 9.11 illustrate the outcome of investigations of these effects. Carbon nanotubes of various thicknesses were grown on Ni substrates. Figure 9.12 shows aligned CNTs with areal densities ranging from 7.5×10^5 to 3×10^8 cm^{-2}.

The outcome of so-called super-growth, which favours the SWCNT variety, is shown as a densely packed aligned array in Figure 9.12.

For revealing structure down to the level of single atoms, TEM analysis is the preferred technique. The high-resolution phase contrast image in Figure 9.13 shows the coaxial tube structure of a MWCNT.

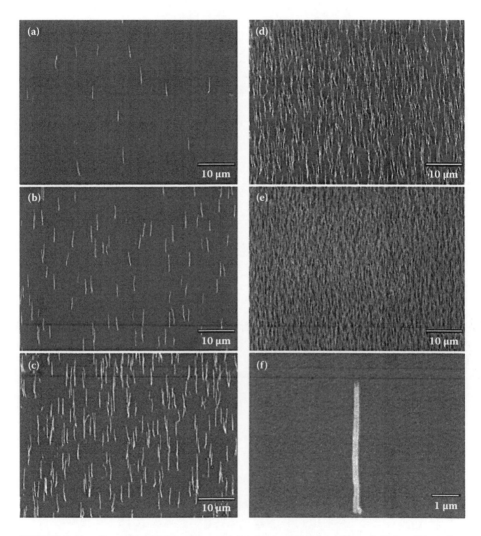

FIGURE 9.11 Growth of CNTs with areal densities of (a) 7.5×10^5 cm^{-2} (b) 2×10^6 cm^{-2} (c) 6×10^6 cm^{-2} (d) 2×10^7 cm^{-2} and (e) 3×10^8 cm^{-2} (f) a single CNT. (From Tu et al., 2002, *Appl. Phys. Lett.*, with permission from American Institute of Physics.)

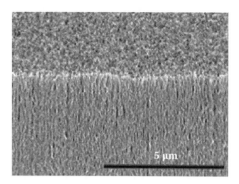

FIGURE 9.12 SEM photo of SWNT forests produced by 'super-growth' on a substrate in the presence of water in the CVD chamber. (From Wikipedia, with permission from Hata et al., 2004.)

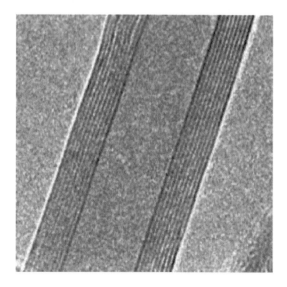

FIGURE 9.13 An HRTEM image showing the structure of an MWCNT. The inner diameter is ca. 10 nm, there are nine nested coaxial walls, and an open inner channel. (From NanoLab Inc., http://www.nano-lab.com/nanotube-image3.html.)

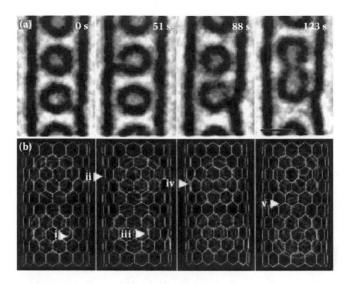

FIGURE 9.14 **(See color insert.)** HRTEM imaging of C_{92} molecules inside an SWCNT; effects of electron irradiation dose. (From Urita et al., 2004, *Nano Lett.*, with permission from the American Chemical Society.)

The HRTEM image in Figure 9.14 shows C_{92} fullerenes within the core space of an SWCNT. The image demonstrates the effect of beam irradiation giving rise to movement of the fullerenes, and of a transient attached to the wall of the tube, as well as to incipient dumbbell formation of two fullerenes within the confining space (Urita et al. 2004).

9.7.1.2 Other Tube Structures

Boron nitride nanotubes (BNNT) have topography and structure analogous to that of CNT, and, accordingly, the same techniques for their characterization can be used. The images below illustrate typical applications of electron-optical methods (see Figure 9.15).

Selected area diffraction of nanostructures is rarely useful, due to the relatively large interaction volume. However, in some cases where the diameter of a multi-wall NT is in the range of 100 nm, then valuable information can be obtained, as illustrated in Figure 9.16.

The characteristic reflections and directions of the tube axes in the diffraction patterns (DPs) are marked. The small arrow in Figure 9.16b indicates the {210} reflection that is forbidden in cylindrical tubes but appears here because of the polygon-like cross-section of the BNNT. Both DPs are characteristic of the preferred zig-zag tube layer orientation found in the graphite-like shells consisting of BN, with their (10,10) directions parallel to the tube axes. Deviations from the ideal zig-zag orientation do not exceed ca. $10°–13°$, as measured by the characteristic semi-angle between the reflection couples in the DPs, originating from the front and back NT sections along the incident TEM electron beam.

FIGURE 9.15 A typical SEM image displaying high yield, pure phase, multi-walled BN nanotubes. The inset on the left shows a representative HRTEM image of an isolated, well-structured, eight-walled BNNT; that on the right gives a comparison between the appearance in bulk of 0.25 g of BNNTs (the white colour of the product is due to non-absorbance of visible light) and conventional black CNTs. (From Golberg et al., 2007, *Adv. Mater.*, with permission from John Wiley and Sons.)

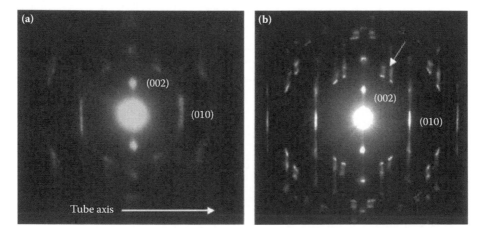

FIGURE 9.16 Electron diffraction patterns (DPs) recorded from individual multi-walled BNNTs. The pattern in (a) is from a tube having a cylinder-like cross-section (Golberg et al., 1999, *J. Appl. Phys.*), while that in (b) is from a tube exhibiting a polygon-like cross-section. (From Golberg et al., 2007, *Adv. Mater.*, with permission from John Wiley and Sons.)

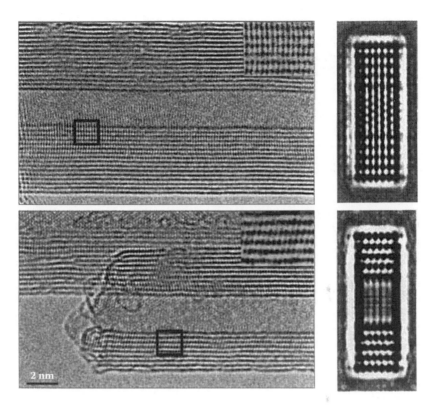

FIGURE 9.17 HRTEM images of two zig-zag BN MWNTs. Distinct, but different, stacking orders are apparent in the marked areas in (upper) and (lower), as highlighted in the insets. Hexagonal- (upper) and rhombohedral-type (lower) stacking (in 12.5° fringe inclinations with respect to the tube axis) were verified by computer-simulations (right panels) for BN MWNTs having the axes strictly parallel to the (10,10) orientations (zig-zag NTs). Note the open tip of a BNNT in the lower image; this feature is common for most BNNTs produced by high-temperature chemical syntheses. (From Golberg et al., 2000, *Appl. Phys. Lett.*, with permission from the American Institute of Physics.)

HRTEM images of MWBNNTs revealing two types of stacking sequence, verifiable by computer simulations, are shown in Figure 9.17.

EELS spectra taken from an individual pure BNNT exhibit core-loss K-edges of B and N at 188 and 401 eV, respectively. The spectra also show that the 1s–p* (left-hand side peaks of the edges) and 1s–r* (right-hand side peaks of the edges) transition features are nearly identical to those of layered BN (hexagonal or rhombohedral) (Gleize et al. 1994). The deduction to be drawn is the existence of an overall sp^2-hybridised structure. On the other hand, it has been reported that the valence electron excitation (so-called plasmon loss) spectra of BNNTs are considerably different from those of layered BN (Terauchi et al. 1998). This may be due to a decrease in the bandgap of 0.6–0.7 eV in BNNTs compared with that in layered BN, due to curvature of the sheet into a nanotube. As far as BN doping is concerned, the ternary BN–C system has attracted most attention because of the marked similarities between layered BN and

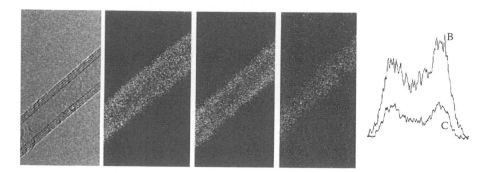

FIGURE 9.18 Elemental distributions in BN–C nanotubes, as revealed by energy-filtered TEM: An HRTEM image is shown in the left panel. The three images in the middle, left to right, demonstrate the homogeneous distribution of all constituent (B, C, and N, respectively) species. Elemental profiles recorded across the homogeneous NT illustrate the expected correlation between the B and C count intensities. (From Gleize et al., 1994, *J. Mater. Sci.*, with kind permission from Springer Science+Business Media.)

C-based materials, and the possibility of producing a range of B–C–N ternary tubes. Spatially resolved EELS and energy-filtered electron microscopy are becoming the most useful experimental techniques for the chemical analysis of C-doped BN and ternary BN–CNTs (e.g., Golberg et al. 2002). Typical results are shown in Figure 9.18.

A series of SEM images, Figure 9.19, shows an ordered array of TiO_2 nanotubes grown on a substrate; the structures were synthesised and investigated in order to determine photocatalytic properties.

The topography of WO_3 tubes is shown in Figure 9.20, as an example of tube structures in the size range ca.100 nm.

The STEM images in Figure 9.21 of MoS_2 MWNTs show examples of tube end capping, and an illustration of the point-to-point resolution that can be obtained with an aberration-corrected instrument operated in the high angle annular dark field (HA-ADF) mode. The principal objective of the analysis was to investigate the effect of end capping on field emission. Image simulation shown in Figure 9.22 validates the interpretation of the images.

9.7.2 ANALYSIS OF NANOTUBES BY SPM

There are some cases where SPM imaging provides insights not readily available with electron optical techniques. For instance, an STM image constitutes an atomically resolved map of the density of states in real space, as illustrated in the case of a chiral CNT in Figure 9.23.

AFM is a user-friendly tool for the visualisation of 3-D topography, with the advantages that specimen preparation is not usually required, that good service in an ambient atmosphere can be provided, and that the specimen is not exposed to ionizing radiation. An example is shown in Figure 9.24 of the configuration adopted for an early demonstration of the ability of a CNT to function as the active element in a FET device.

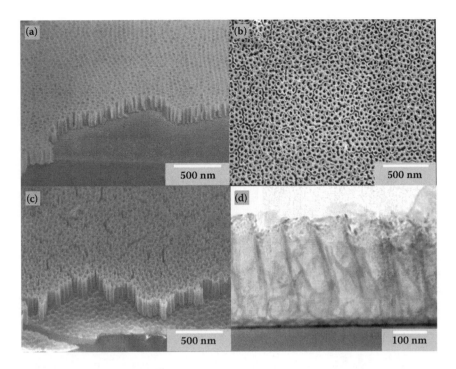

FIGURE 9.19 Imaging by SEM and TEM of a free-standing porous anodic alumina (PAA) template and TiO_2 nanotube arrays (NTAs) on glass substrates. (A) SEM image of a free-standing PAA template adhering to a glass substrate, showing hexagonally packed pores with an average diameter of 65 nm and interpore spacing of 110 nm. (B) Top-view SEM image of dense and vertically well-aligned TiO_2 NTAs on a glass substrate after being released from the PAA template. The NTAs are 200 nm in height with a wall thickness of 15 nm. (C) Oblique SEM image showing TiO_2 NTAs intimately attached to a glass substrate even after cleavage of the substrate. The NTAs are assumed to be chemically bonded to the substrate. (D) Cross-sectional TEM image of TiO_2 NTAs on a glass substrate that has been functionalized with Pd nanoparticles acting as a catalyst for photocatalytic reactions. (From Tan et al., 2010, *ACS Appl. Mat. Interfaces*, with permission from the American Chemical Society.)

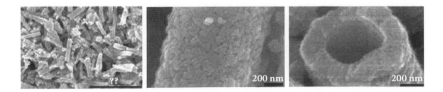

FIGURE 9.20 SEM images of WO_3 nanotubes. (From *AIST Today*, 2009.)

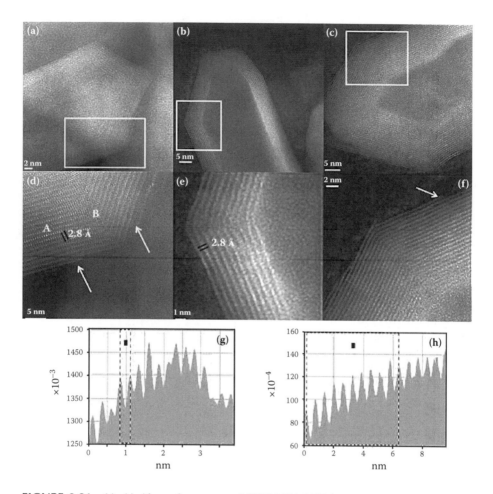

FIGURE 9.21 (a)–(c) Aberration-corrected STEM-HAADF images recorded from MoS_2 nanotubes showing conical and facetted end capping in (a) and (b), The higher resolution images in (d)–(f) reveal details of the layer structures near the end caps of the nanotubes shown in (a), (b) and (c) (marked by the boxes), In (g) is a line profile of the Mo-Mo spacing of the nanotube layers shown in (e), which shows it to be close to 0.28 nm, while in (h) a line profile of the nanotube layers shown in (f) indicates an interlayer spacing of ca. 0.63 nm. The white arrows in (d) and (f) mark the corners of the nanotubes, where the last atomic MoS_2 layer appears to be discontinuous. (From Deepak et al., 2010, *Nanoscale*, reproduced with permission from the Royal Society of Chemistry.)

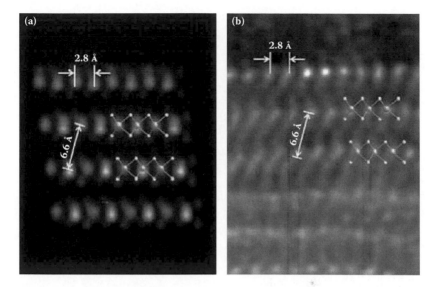

FIGURE 9.22 (a) Simulated image of a segment of the multi-wall tube structure imaged by STEM in Figure 9.21. The resolution and contrast are sufficient to identify the S columns, as shown by the superposition of a ball-and-stick model. (b) Segment of an aberration-corrected STEM micrograph, in which the S columns are also identifiable. The interlayer distance was measured as 0.66 nm, in agreement with the spacing for the planar compound. (From Deepak et al., 2010, *Nanoscale*, reproduced with permission from the Royal Society of Chemistry.)

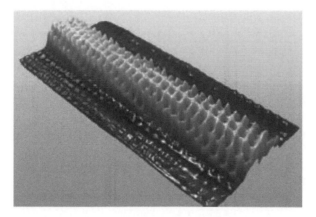

FIGURE 9.23 (**See color insert.**) STM image of a chiral CNT; the diameter of the tube was ca. 1 nm. (From the Image Cees Dekker Group at TU Delft, The Netherlands, reproduced with permission from the author.)

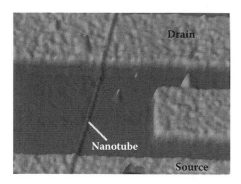

FIGURE 9.24 **(See color insert.)** An AFM image showing an FET device constructed by bridging source and drain electrodes by a single CNT. The SiO_2 substrate constitutes the gate dielectric. (From Martel et al., 1998, *Appl. Phys. Lett.*, with permission from the American Institute of Physics.)

Another example of AFM imaging of a CNT-based device (intended to test quantum effects in an SWCNT structure) (Inzani 2011) is shown in Figure 9.25. Single- or few-wall CNTs were grown by a CVD method (similar to that of Liu et al. 2010). The active part of the device was subsequently defined by lithographic deposition of source and drain electrode pads, see SEM image in Figure 9.25. It was verified by AFM analysis that the CNT bridging the electrode gap was of the single-wall variety, and that it was continuous.

9.7.3 RAMAN SPECTROSCOPY OF CNTs

Even though conventional Raman microprobe spectrometers cannot provide laser spots smaller than ca. 0.75 μm in diameter, it should be remembered that the spectra actually arise from photon interaction with the vibrations of individual chemical bonds. This intrinsic 'nanoprobing' makes Raman spectroscopy sensitive to short-range structure. Also, since almost no sample preparation is necessary, it is often possible and cost-effective to use the technique as a basic nanoprobe.

Raman spectroscopy has been used to investigate carbon-containing materials for many years (e.g., Lewis and Edwards 2001). The allotropes of carbon provide a rich vein for research into relating spectral features to molecular and electronic structure, and CNT vibrational properties have been widely investigated by Raman spectroscopy (Castiglioni et al. 2006; Shimada et al. 2005; Dresselhaus et al. 2005; Brar et al. 2002; Rao et al. 1998; Dresselhaus, Dresselhaus, and Jorio 2002; Pfeiffer et al. 2002).

Characteristics of CNT, such as crystalline quality and phase purity, can be monitored using quality indicators derived from the Raman spectra. The Raman spectrum of a CNT consists of bands in a low-energy range, due to radial breathing modes (RBM), to dispersive or D bands around 1300–1400 cm^{-1}, and to so-called G bands near 1600 cm^{-1}, due to tangential modes. The RBM modes are sensitive to the diameter and chirality/helicity of small diameter tubes, and depend also on the excitation wavelength, with narrow resonance profiles (<100 meV) and a small broadening and energy shift (when the CNTs are in bundles) (Donato et al. 2007).

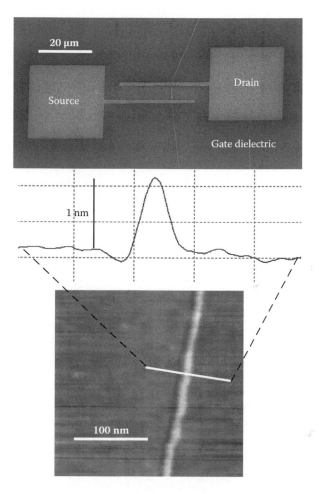

FIGURE 9.25 (**See color insert.**) Illustration of SEM and AFM characterization of a CNT-based device. The SEM image (top) shows the lay-out of the device, while the middle AFM line trace, taken from the AFM tapping mode image (bottom), demonstrates that the CNT was single wall. (From Inzani, 2011, unpublished data, University of Oxford, with permission.)

The wave numbers of the RBM modes, ν_{RBM}, depend inversely on the tube diameter d, according to the relationship

$$\nu_{RBM} = A/d + B \qquad (9.10)$$

However, a variety of values have been reported for the constants A and B, the variations often being attributed to environmental effects, such as whether the SWCNTs are present as individual tubes covered with a surfactant, or isolated on a substrate, or in the form of bundles (Graupner 2007). But for isolated SWCNTs, where the chiral vector and therefore the diameter have been determined independently by electron diffraction, values of A and B of 204 cm^{-1} and 27 cm^{-1} have been established (Graupner 2007; Puech et al. 2007; Gui et al. 2003; Fantini et al. 2005).

The tangential modes are the most intense high-energy modes of SWCNTs and form the so-called G band, observed near 1600 cm^{-1}, which is close to the position of the G mode in graphite. It consists of two bands, each of symmetric line shape, one at 1590 cm^{-1} (G$^+$), and another at ca. 1572 cm^{-1} (G$^-$), for tubes with diameters of 1.4 nm. The position of the G$^+$ mode is not sensitive to the tube diameter. The precise position of the G$^-$ band depends on (C/d^2), where C is a constant depending on the semiconducting or metallic character, and d is the diameter. Coupling with valence electrons in metallic CNTs has a strong influence on the band shape and intensity of the G bands (Fantini et al. 2005). For multi-wall carbon nanotubes (MWCNTs), the number of walls, and the presence of faceted graphitic particles, complicate the G-band intensity and band shape. On the higher wave number side of the G band, a shoulder, due to the so-called D' band, is also sometimes observed (Meyer et al. 2005).

Probably the most discussed mode for the characterization of functionalised SWCNTs is the D band, observable at 1300–1400 cm^{-1}, and related to defect-induced double-resonant scattering processes involving elastic scattering of electrons by structural defects. In their early work on the Raman spectroscopy of graphite, it was shown by Tuinstra and Koenig (1970) that the intensity of this band scales linearly with the inverse of the crystallite size, and its appearance was interpreted as being due to a breakdown of the k-selection rule. However, for graphite and SWCNTs, the position of this mode, as well as its intensity, with respect to the G band, depends on the wavelength of the exciting laser. The D band also displays a dependence on the diameter of a SWCNT, which has been interpreted as a double-resonance process in which not only one of the direct, k-conserving, electronic transitions, but also the emission of the phonon, is a resonant process. In contrast to single-resonant Raman scattering, in which only phonons around the centre of the Brillouin zone ($q = 0$) are excited, the phonons that contribute to the D band exhibit a non-negligible q-vector. Overall, k-conservation for the Raman scattering process is fulfilled by elastic defect scattering. Taking the electronic as well as the phonon dispersion relation into account, this double-resonance theory is able to explain the intensity of the D band as a function of laser energy, as well as the fine structure observed in the D-band spectra (Tuinstra and Koenig 1970). The ratio of peak heights of the G to D bands can also be used to assess the quality of the SWCNTs; the ratio is taken as an indicator of the relative proportion of highly ordered graphitic carbon to disordered or amorphous carbon. High quality CNTs can have a ratio greater than 100.

The overtone of the D band, the G' band (sometimes called the 'D* band'), observed at 2600–2800 cm^{-1}, does not require explanation by defect scattering, since two phonons with q and $-q$ are excited. This mode is, therefore, observed independently of defect concentration. In fact, the intensity ratio I_D/I_G is thought by some authors (Maultzsch et al. 2002; Murphy, Papakonstantinou, and Okpalogu 2006) to be a sensitive measure of the defect concentration in carbon nanotubes. An assignment of additional modes visible in the Raman spectra of SWCNTs can be found in the review literature (Saito et al. 2003; Dresselhaus et al. 2005).

Figure 9.26 shows a typical spectrum from an SWCNT, excited by red light of wavelength 633 nm from a HeNe laser, in which the RBM, D and G bands, as well as the two-phonon overtone band at ca. 2600 cm^{-1}, can be seen (Graupner 2007), while

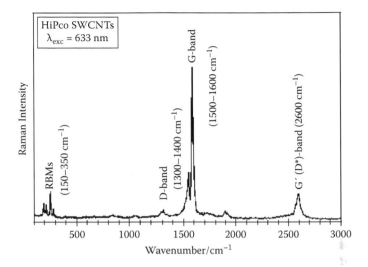

FIGURE 9.26 Raman spectrum from an SWCNT, showing the RBM, D, and G bands and the overtone band at ca. 2600 cm⁻¹. The exciting laser wavelength was 633 nm. (From Popov and Lambin, 2006, *Phys. Rev. B*, with permission from American Physical Society.)

Figure 9.27 shows Raman spectra recorded from purified SWCNTs excited by five different wavelengths (Rao et al. 1998).

Raman scattering can also be used as a probe of doped SWNT material (Gai et al. 2004). Figure 9.28 shows the spectral region from 100~1850 cm⁻¹, which contains the first-order features of the SWNT Raman spectrum. The significant feature in the spectra, related to the concentration of the boron dopant, is the D band. It is clear that the structure remains substantially intact up to a nominal 3 at.% concentration of boron. However, when the boron concentration is increased beyond that value, structural disorder is induced giving rise to an increasing intensity in the D band due to a double resonance effect. The result is in agreement with a study (Hishiyama, Irumano, and Kaburagi 2001) of boron doping in highly oriented graphite films.

The polarization dependence of Raman spectroscopy can provide additional information. An exhaustive study of aligned SWCNTs has been carried out by Murakami et al. (2005). The geometry is shown in Figure 9.29. The emphasis was on the dependence of the RBM modes on the directions of propagation and polarisation, and on laser wavelength and tube diameter (0.8–3 nm). Representative results are shown in Figure 9.30.

The measured peaks of the radial breathing mode (RBM) stimulated by incident photons of energy 1.96, 2.41, and 2.54 eV, (the photonic energy is a significant variable because it affects the extent to which there will be contributions to the Raman spectra from excited state configurations), can be categorized into two groups. One group of peaks (at ca. 120 and 150 cm⁻¹) is dominant for perpendicular polarisation while another group (at ca. 160 cm⁻¹) is dominant for light polarized parallel to the SWNT axis. The results are consistent with peaks dominant for parallel and perpendicular configurations being associated with $\Delta\mu = 0$ and $\Delta\mu = \pm1$ electronic excitations, respectively ($\Delta\mu$ refers to selection rules based on the polarisation tensor).

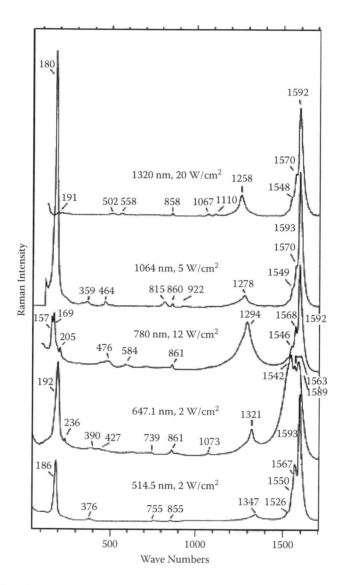

FIGURE 9.27 Room temperature Raman spectra from purified SWCNTs excited at five different laser wavelengths and power densities. (From Rao et al., 1998, *Thin Solid Films*, with permission from Elsevier.)

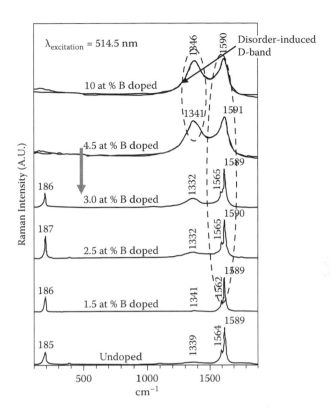

FIGURE 9.28 First-order Raman spectra of SWCNTs doped with different boron concentrations. (From Gai et al., 2004, *J. Mat. Chem.*, with permission to reproduce from the Royal Society of Chemistry.)

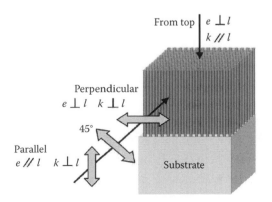

FIGURE 9.29 Schematic description of the relationships between the direction of laser propagation (k), the direction of laser polarisation (e), and the direction of SWNT axis (l). (From Murakami et al., 2005, *Phys. Rev. B*, with kind permission from Dr. S. Murakami.)

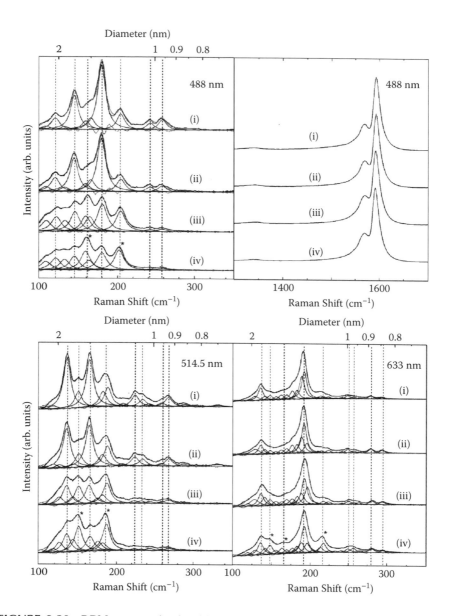

FIGURE 9.30 RBM spectra stimulated by lasers of wavelengths 488, 514.5, and 633 nm for different incident configurations. The directions of polarisation of the spectra were: (i) top, (ii) perpendicular, (iii) 45°, and (iv) parallel, configurations. G-band spectra taken at 488 nm are also shown. The RBM spectra were normalised to the corresponding G+ heights and resolved into Lorentzian components by maintaining the FWHM values the same within each spectrum. (From Murakami et al., 2005, *Phys. Rev. B*, with kind permission from Dr. S. Murakami.)

The grouping behaviour of RBM peaks is also a function of changes in incident laser power. Finally, the spectra exhibit the expected dependence of the RBM modes on the tube diameter (the expected positions of the main RBM modes for tubes of different diameters are shown at the tops of the graphs). The work summarised in Figure 9.30 has been updated with additional results and an enhanced interpretation (Zhang et al. 2010).

The analysis of single CNTs with conventional Raman spectroscopy is of limited use, even with a confocal microprobe instrument, due to the diffraction limitation of the spatial resolution. The count rate can be improved in the SERS mode, and there are numerous examples of that in the literature. More recently, aperture-less scanning near-field optical microscopy or near-field scanning optical microscopy (SNOM/NSOM) has become available; accordingly, tip-enhanced Raman spectroscopy (TERS) has become a practical and effective way of performing spectroscopic imaging with spatial resolution in the tens of nm range (e.g., Renishaw 2008; Cançado, Hartschuh, and Novotny 2009).

9.7.4 CHARACTERIZATION OF ELECTRONIC STRUCTURE

9.7.4.1 EELS

EELS attachments to HRTEM instruments are now widely available. If EELS is deployed in combination with an energy-filtered electron beam, then the spectral resolution is typically 0.25 eV, and chemical information can be obtained. As well, useful information can be obtained from plasmon structures in the spectra. However, in the case of CNT analysis, the chemical shifts are generally insufficient to extract useful chemical information. Accordingly, in most cases EELS is used as a high spatial resolution analytical tool in combination with HRTEM imaging.

Results from an investigation of BN-coated MWCNTs are shown in Figures 9.31 and 9.32 (Chen, Ye, and Gogotsi 2004). Two kinds of MWCNT constituted the starting material: (1) pyrolytically stripped (purified by pyrolysis) CNT (PS-CNTs), and (2) heat-treated CNT (HT-CNTs). The average diameter (ca. 100 nm) and wall thickness (15–20 nm) were similar for both types. The MWCNTs were subsequently clad with BN in a two-step process, commencing with boric acid infiltration and followed by nitridation (Chen et al. 2003).

9.7.4.2 Luminescence Spectroscopy

It has been shown that SWCNTs exhibit photoluminescence, at their near-IR E_{11} transition wavelengths (O'Connell et al. 2002). With the subsequent successful assignment of the E_{11} and E_{22} (Mann et al. 2006) spectral transitions, (the notation refers to transitions between particular discrete energy levels), to specific (m,n) chiral structures (Bachilo et al. 2002), optical spectroscopy has emerged as a useful tool for the correlation of crystal structure with the electronic structure of SWNTs (Weisman 2009). The method has become known as fluorimetry.

Results from fluorimetric analysis are shown as a pseudo-3D plot in Figure 9.33, where the peak heights correspond to intensity of emitted radiation. The two other axes represent excitation and emission wavelengths. The peaks dispersed in the x-y

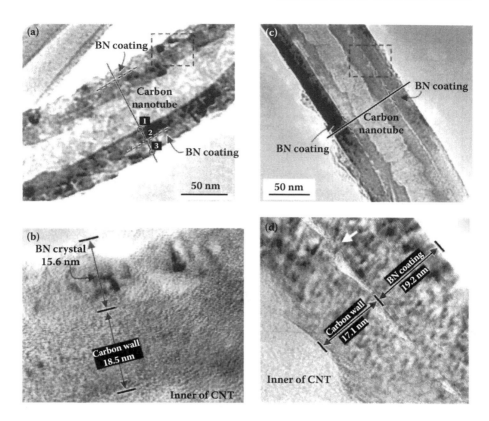

FIGURE 9.31 TEM images of BN-coated PS-CNTs (a, b) and HT-CNTs (c, d). The solid lines in (a) and (c) indicate the paths of the line-scan EELS analyses shown in Figure 9.32; (b) and (d) are HRTEM images of the framed regions in (a) and (c), respectively. (From Chen et al., 2004, *J. Am. Ceram. Soc.*, with permission from John Wiley and Sons.)

plane indicate the excitation in the E_{22} range between 500 and 800 nm. Each peak arises from a distinct (m,n) chirality of SWNT that corresponds to semiconducting band structure. The unique E_{11} and E_{22} transition energies for a particular chirality can be determined from the wavelength coordinates.

The data in Figure 9.33 are shown as a false-colour intensity map in Figure 9.34, and in Figure 9.35 a plot of E_{22}/E_{11} as a function of excitation wavelength. The data points appear to form a systematic pattern. These data reflect the systematic pattern of nanotube structures present in the sample.

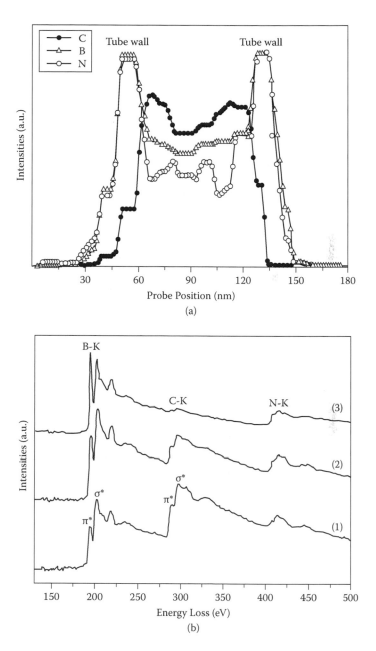

FIGURE 9.32 EELS analysis of a BN-coated PS-CNT: (a) normalized concentration profiles of B, C, and N across the nanotube (along the line shown in Figure 9.31(a). (b) The evolution of the EELS spectra from the inner tube wall (point 1 in Figure 9.31(a)) to the edge of the BN coating (point 3) shows the near-edge fine structure of B, C, and N. (From Chen et al., 2004, *J. Am. Ceram. Soc.*, with permission from John Wiley and Sons.)

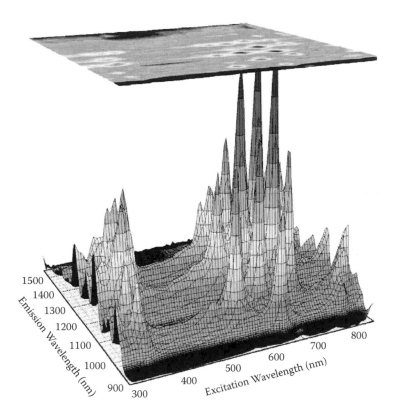

FIGURE 9.33 Pseudo-3D plot of the fluorimetric response of a sample of SWCNTs with a range of chiralities. The peaks represent semiconducting CNTs, in which the range of band gaps is dispersed in the x-y plane according to excitation and emission wavelengths. (From Weisman, 2008, *Contemporary Concepts of Condensed Matter Science*, with permission from Elsevier.)

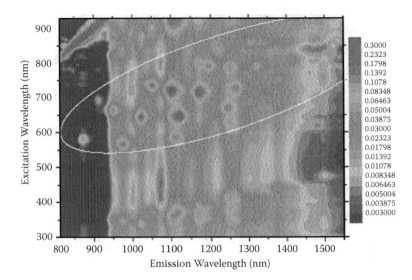

FIGURE 9.34 (**See color insert.**) False-colour contour plot of the data in Figure 9.33 shows the precise wavelengths for each peak. (From Weisman, 2008, *Contemporary Concepts of Condensed Matter Science*, with permission from Elsevier.)

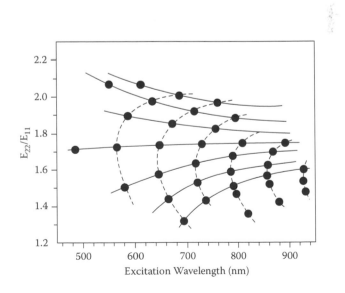

FIGURE 9.35 An apparently systematic relationship for the data in Figures 9.33 and 9.34. (From Weisman, 2008, *Contemporary Concepts of Condensed Matter Science*, with permission from Elsevier.)

REFERENCES

AIST, 2008, press release, August 4.

AIST Today, 2009, 8(11), p. 70.

Arepalli, S., Nikolaev, P., Gorelik, O., and VHadjiev, V. G., 2004, Protocol for the Characterization of Single-Wall Carbon Nanotube Material Quality (NASA JSC protocol), *Carbon*, 42, 1783.

Bachilo, S. M., Strano, M. S., Kittrell, C., Hauge, R. H., Smalley, R. E., and Weisman, R. B., 2002, *Science*, 298, 2361.

Berciaud, S., Cognet, L., Poulin, P., Weisman, R. B., and Lounis, B., 2007, *Nano Lett.*, 7, 1203.

Bethune, D. S., Kiang, C. H., de Vries, M. S., Gorman, G., Savoy, R., Vazquez, J., and Beyers, R., 1993, *Nature*, 363, 605.

Blasé, X., Rubio, A., Louie, S. G., and Cohen, M. L., 1994, *Europhys. Lett.*, 28, 335.

Bonard, J. M., Kind, H., Stöckli, T., and Nilsson, L. O., 2001, *Solid-State Electronics*, 45, 893.

Brar, V. W., Samsonidze, G. G., and Dresselhaus, M. S., 2002, *Phys. Rev. B*, 66, 155418.

Cassell, A. M., Franklin, N. R., Tombler, T. W., Chan, E. M., Han, J., and Dai, H., 1999, *J. Am. Chem. Soc.*, 121, 7975.

Castiglioni, C., Negri, F., Tommasini, M. Di Donato, E., and Zerbi, G., 2006, In *Carbon: The Future Material for Advanced Technology Applications*, G. Messina and S. Santangelo, (eds.), Heidelberg: Springer Series Topics in Applied Physics 100.

Chopra, N. G., Luyken, R. J., Cherrey, K., Crespi, V. H., Cohen, M. L., Louie, S. G., and Zettl, A., 1995, *Science*, 269, 966.

Cançado, L. G., Hartschuh, A., and Novotny, L., 2009, *J. Raman Spectrosc.*, 40, 1420.

Chen, L., Ye, H., and Gogotsi, Y., 2004, *J. Am. Ceram. Soc.*, 87, 147.

Chen. L., Ye, H., Gogotsi, Y., and McNallan, M. J., 2003, *J. Am. Ceram. Soc.*, 86, 1830.

Dai, H., Kong, J., Zhou, C., Franklin, N., Tombler, T., Cassell, A., Fan, S., and Chapline, M. 1999, *J. Phys. Chem. B*, 103, 11246.

Deepak, F. L., Mayoral, A., Steveson, A. J., Mejıa-Rosales, S., Blom, D. A., and Jose-Yacaman, M., 2010, *Nanoscale*, 2, 2286.

Donato, M. G., Messina, G., Santangelo, S., Galvagno, S., Milone, C., and Pistone, A., 2007, *J. Phys. Conf. Ser.*, 61, 931.

Dresselhaus, M. S., Dresselhaus, G., and Jorio, A., 2002, *Carbon*, 40, 2043.

Dresselhaus, M. S., Dresselhaus, G., and Saito, R., 1992, *Phys. Rev. B*, 45, 6234.

Dresselhaus, M. S., Dresselhaus, G., Saito, R., and Jorio, A., 2005, *Phys. Rep.*, 409, 47.

Fantini, J. A., Pimenta, M. A., Capaz, R. B., Samsonidze, G. G., Dresselhaus, G., Dresselhaus, M. S., Jiang, J., Kobayashi, N., Gruneis A., and Saito, R., 2005, *Phys. Rev. B*, 71, 075401.

Farhat, S., 2001, *J. Chem. Phys.*, 115, 6752.

Franklin, N. and Dai, H., 2000, *Adv. Mater.*, 12, 890.

Gai P. L., Stephan O., McGuire K., Rao A. M., Dresselhaus M. S., Dresselhaus G., and Colliex C., 2004, *J. Mat. Chem.*, 14, 669.

Gleize, P., Schouler, M. C., Gadelle, P., and Caillet, M., 1994, *J. Mater. Sci.*, 29, 1575.

Golberg, D., Bai, X. D., Mitome, M., Tang, C., Zhi, C. Y., and Bando, Y, 2007, *Acta Mater.*, 55, 1293.

Golberg, D., Bando, Y., Bourgeois, L., Kurashima, K., and Sato, T., 2000, *Appl. Phys. Lett.*, 77, 1979.

Golberg, D., Bando, Y., Mitome, M., Kurashima, K., Grobert, N., Reyes-Reyes, M., Terrones, H., and Terrones, M., 2002, *Chem. Phys. Lett.*, 360, 1.

Golberg, D., Bando, Y., Tang, C., and Zhi, C., 2007, *Adv. Mater.*, 19, 2413.

Golberg, D., Han, W., Bando, Y., Kurashima, K., and Sato, T., 1999, *J. Appl. Phys.*, 86, 2364.

Gong, D., Grimes, C. A., Varghese, O. K., Hu, W., Singh, R. S., Chen, Z., and Dickey, E. C., 2001. *J. Mater. Res.*, 16, 3331.

Graupner, R., 2007, *J. Raman Spectrosc.*, 38, 673.

Gui, S., Canet, R., Derre, A., Couzi, M., and Delhaes, P., 2003, *Carbon*, 41, 41.

Hamada, N., Sawada, S., and Oshiyama, A., 1992, *Phys. Rev. Lett.*, 68, 1579.

Hata, K., Futaba, D. N., Mizuno, K., Namai, T., Yumura, M., and Iijima, S., 2004, *Science*, 306, 1362.

Hishiyama Y., Irumano H., and Kaburagi Y., 2001, *Phys. Rev. B*, 63, 245406.

Iijima, S., 1991, *Nature,* 354, 56.

Inzani, K., 2011, Part II project, Oxford University (unpublished data).

Jishi, R. A., Inomata, D., Nakao, K., Dresselhaus, M. S., and Dresselhaus, G. 1994, *J. Phys. Soc. Jap.*, 63, 2252.

Kim, J. M. Choi, W. B., Lee, N. S., and Jung, J. E., 2000, *Diamond Relat. Materials*, 9, 1184.

Lee, N. S., Chung, D. S., Han, J. T., Kang, J. H., Choi, Y. S., Kim, H. Y., Park, S. H., Jin, Y. W., Yi, W. K., Yun, M. J., Jung, J. E., Lee, C. J., You, J. H., Jo, S. H., Lee, C. G., and Kim, J. M., 2001, *Diamond Relat. Materials*, 10, 265.

Lewis I. R. and Edwards, H. G. M., 2001, In *Handbook of Raman Spectroscopy*, Boca Raton, FL: CRC Press. The volume contains more than 1,000 references to Raman spectroscopy of elemental carbon-based materials.

Liu, Z., Jiao, L., Yao, Y., Xian, X., and Zhang, J., 2010, *Adv. Materials*, 22, 2285.

Mann, D., Kato, Y. K., Kinkhabwala, A., Pop, E., Cao, J., Wang, X., Zhang, L., Wang, Q., Guo, J., and Dai, H., 2006, *Nature Nanotechnology*, 2, 33.

Martel, R., Schmidt, T., Shea, H. R., Hertel, T., and Avouris, P., 1998, *Appl. Phys. Lett.*, 73, 2447.

Maultzsch, J., Reich, S., Thomsen, C., Webster, S., Czerw, R., Carroll, D. L., Vieira, S. M. C., Birkett, P. R., and Rego, C. A., 2002, 81, 2647.

Meyer, J. C., Paillet, M., Michel, T., Moreac, A., Neumann, A., Duesberg, G., Roth S., and Sauvajol, J. L., 2005, *Phys. Rev. Lett.*, 95, 217401.

Milne, W. I., Teo, K. B. K., Amaratunga, G. A. J., Legagneux, P., Gangloff, L., Schnell, J. P., Semet, V., Thien Binh, V., and Groening, O., 2003, *J. Mater. Chem.,* 14, 933.

Mintmire, J. W., Dunlap, B. I., and White, C. T., 1992, *Phys. Rev. Lett.*, 68, 631.

Murakami, Y., Chiashi, S., Einarsson, E., and Maruyama, S., 2005, *Phys. Rev. B*, 71: 085403.

Murphy, H., Papakonstantinou, P., and Okpalogu, T.I.T., 2006, *J. Vac. Sci. Technol. B*, 24, 715.

O'Connell, M. J., Bachilo, S. M., Huffman, C. B., Moore, V., Strano, M. S., Haroz, E., Rialon, K., Boul, P. J., Noon, W. H., Kittrell, C., Ma, J., Hauge, R. H., Weisman, R. B., and Smalley, R. E., 2002, *Science*, 297, 593.

Odom, T. W., Huang, J. L., Kim, P., and Lieber, C. M., 1998, *Nature*, 391, 62.

Park, S., Kim, H. C., Yum, M. H., Yang, J. H., Park, C. Y., Chun, K. J., and Eom, B, 2008, *Nanotechnology,* 19, 445304.

Park, C. K., Kim, J. P., Yun, S. J., Lee, S. H., and Park, J. S., 2007, *Thin Solid Films*, 516, 304.

Pfeiffer, R., Kuzmany, H., Pank W., and Pichler, T., 2002, *Diamond Relat. Mater.*, 11, 957.

Popov, V., 2004, *New J. Phys.*, 6, 1.

Popov, V. N. and Lambin, P., 2006, *Phys. Rev. B*, 73, 85407.

Pouch, J. J. and Alterovitz, A., 1990, *Synthesis and Properties of Boron Nitride*, Vol. 54 and 55, Zürich: Trans. Tech. Publications.

Puech, P., Flahaut, E., Bassil, A., Juffmann, T., Beuneu, F., and Basca, W. S., 2007, *J. Raman Spectrosc.*, 38, 714.

Rao, A. M., Bandow, S., Ritcher E., and Eklund, P. C., 1998, *Thin Solid Films,* 331, 141.

Reich, S. and Ordejón, C. T. S. P., 2002, *Phys. Rev. B*, 65, 155411.

Renishaw, 2008, A Study of Single Wall Carbon Nanotubes, Application Note from Spectroscopy Product Development, *Issue* 1.2.

Rothschild, A., Frey, G. L., Homyonfer, M., Tenne, R., and Rappaport, M., 1999, *Mat. Res. Innovat.*, 3, 145.

Saito, R., Dresselhaus, G., and Dresselhaus, M., 1998, *Physical Properties of Carbon Nanotubes*, London: Imperial College Press.

Saito, R., Gruneis, A, Samsonidze, G. G., Brar, V. W., Dresselhaus, G., Dresselhaus, M. S., Jorio, A., Cancado, L. G., Fantini, C., Pimenta, M. A., and Filho, A. G. S., 2003, *New J. Phys.*, 5, 1571.

Shao, L., Tobias, G., Salzmann, C. G., Ballesteros, B., Hong, S. Y., Crossley, A., Davis, B. G., and Green, M. L. H., 2007, *Chem. Comm.*, 47, 5090.

Shimada, T., Sugai, T., Fantini, C., Souza M., and Cançado, L. G., 2005, *Carbon*, 43, 1049.

Tan, L. K., Kumar, M. K., An, W. W., and Gao, H., 2010, *ACS Appl. Mat. Interfaces*, 2, 498.

Terauchi, M., Tanaka, M., Matsumoto, M., and Saito, Y., 1998, *J. Electron Microsc.*, 47, 319.

Thess, A., Lee, R., Nikolaev, P., Dai, H. J., Petit, P., Robert, J., Xu, C. H., Lee, Y. H., Kim, S. G., Rinzler, A. G., Colbert, D. T., Scuseria, G. E., Tomanek, D., Fischer, J. E., and Smalley, R. E., 1996, *Science*, 273, 483.

Tobias, G., Shao, L., Salzmann, C. G., Huh, Y., and Green, M. L. H., 2006, *J. Phys. Chem. B*, 110, 22318.

Tu, Y., Huang, Z. P., Wang, D. Z., Wen, J. G., and Ren, Z. F., 2002, *Appl. Phys. Lett.,* 80, 21.

Tuinstra, F. and Koenig, J. L., 1970, *J. Chem. Phys.*, 53, 1126.

Urita, K., Sato, Y., Suenaga, K., Gloter, A., Hasimoto, A., Ishida, M., Shimada, T., Shinohara, T., and Iijima, S., 2004, *Nano Lett.*, 4, 2451.

Wildöer, J. W. G., Venema, L. C., Rinzler, A. G., Smalley, R. E., and Dekker, C., 1998, *Nature*, 391, 59.

Weisman, R. B., 2008, *Contemporary Concepts of Condensed Matter Science*, 3, 1.

————, 2009, Simplifying Carbon Nanotube Identification, *The Industrial Physicist*, 24.

Zhang, Z., Einarsson, E., Murakami, Y., Miyauchi, Y., and Maruyama, S., 2010, *Phys. Rev. B*, 81, 165442.

Zólyomi, V. and Kürti, J., 2004, *Phys. Rev. B*, 70, 085403.

10 Nanowires

10.1 INTRODUCTION

The designation '*nanowire*' can be used to describe an object with a large aspect ratio and a diameter in the range 1–100 nm. Such a diameter puts the radial dimension of a nanowire at, or below, the characteristic length scales of various interesting and fundamental solid-state phenomena, such as the exciton Bohr radius, the wavelength of light, the phonon mean free path, the critical size of magnetic domains, and so forth. Accordingly, many physical properties of nanowires of metals, alloys, and semiconductors are significantly altered, when compared with the corresponding bulk properties. As well, the large surface-to-volume ratio enhances the relative importance of surface-specific structural and chemical properties such as chemical reactivity. The two-dimensional confinement endows nanowires with properties that are uniquely different from those of the corresponding bulk material, while supporting thermal and electronic transport in the long dimension. Therein are the reasons for nanowires being of great interest to the basic research community.

The large aspect ratio of nanowires is relevant to their technological application. The one unconstrained dimension permits electrical and thermal conduction over microscopic distances. Indeed, 1-D structures are the lowest dimension objects that can be used for electrical and thermal transport. Thus, nanowires may constitute ideal materials for the manufacture of next-generation nanoscale solid-state devices.

A chronological summary of progress in the field can be found in the review literature (Hu et al. 1999; Xia et al. 2003; Hochbaum and Yang 2010; Schmidt et al. 2010, and references therein).

10.2 SYNTHESIS ROUTES

The most widely used synthesis routes are shown schematically in Figure 10.1 (Xia et al. 2003); those authors also provide an excellent overview and description of the broad range of processes, and the process conditions, that can be exploited for the production of nanowires. It is instructive to consider the vapour-liquid-solid (VLS) method in some detail, and nanowire components such as Si and Ge, in order to illustrate the principal techniques and methods used for their characterization. The relative topicality of Si and Ge is due to the fact that the majority of intended applications are conditional on the presence of semiconducting electrical properties.

10.2.1 OVERVIEW OF THE VLS PROCESS

The VLS process dates back to the 1960s (Wagner and Ellis 1964), and remains the most widely used synthesis route. The application of the VLS, and related processes, to the nucleation and growth of Si nanowires on substrates has been the subject of

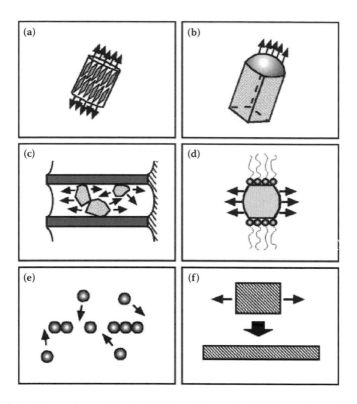

FIGURE 10.1 The various strategies that have been adopted for synthesis of nanowires. (a) Anisotropic growth inherent in the crystal structure. (b) 1-D growth from a liquid drop confined at a nucleation site by a vapour-liquid-solid (VLS) process. (c) I-D growth within an open template channel. (d) Kinetic control by a capping/passivation agent. (e) Self-assembly of 0-D particles. (f) Reduction of diameter by preferentially focussed ion beam erosion or directional etching. (From Xia et al., 2003, *Adv. Materials*, with permission from John Wiley and Sons.)

a recent review (Schmidt et al. 2010). The process is illustrated schematically in Figure 10.2, and a typical outcome is shown (Figure 10.3).

The progressive steps in the VLS process are shown in Figure 10.4. The phase diagram for the Au-Si binary alloy in Figure 10.3 is important because the melting point of the alloy depends on its composition. The lowest melting point of ca. 363°C corresponds to a composition $Si_{19}Au_{81}$. For comparison, the melting points of pure Au and Si are ca. 1050 and 1460°C, respectively. When the temperature of Au in the presence of a source of Si, (such as a film of Au, or dispersed Au seed particles on an Si substrate), is raised to ca. 363°C, liquid droplets of Au-Si will form. If the droplets are then exposed to a gaseous Si precursor, such as SiH_4, the silane will crack at the surfaces of the droplets, thus making additional Si available for solid solution in the droplet. The amount of Si in solution in the droplets will be limited according to the phase diagram. The excess of Si from the gas phase must therefore be disposed of, with the result that solid Si nucleates and crystallises as a wire at the

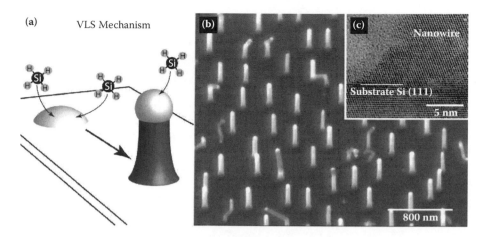

FIGURE 10.2   a) A schematic illustration of the VLS nanowire growth process (Schmidt et al., 2010, *Chem. Rev.*); (b) SEM image of Si nanowires grown epitaxially on a Si (111) substrate; (c) HRTEM image showing the interface region between the nanowire and the substrate. Near-perfect epitaxy is apparent, and curvature of the nanowire pillar at the junction with the substrate is confirmed ((b) and (c)). (Adapted from Schmidt et al., 2005, *Z. Metallkd.*)

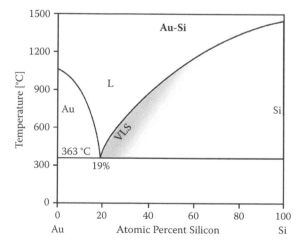

FIGURE 10.3   Graph of temperature versus alloy composition, illustrating the conditions for alloy formation, nucleation, and growth of a Si nanowire. (Adapted from Schmidt et al., 2010, *Chem. Rev.*, with permission from American Chemical Society.)

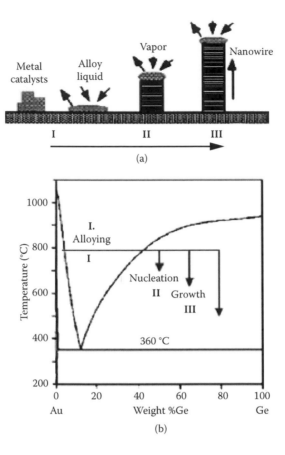

(a)

(b)

FIGURE 10.4 Schematic of successive steps in the VLS process, beginning with the seed metal catalyst particle (Au in this particular case), followed by formation of an intermetallic liquid drop. (From Xia et al., 2003, *Adv. Materials*, with permission from John Wiley and Sons.)

droplet-to-wire interface. The process is shown schematically in Figure 10.4, where the three regimes of alloying, nucleation, and growth are shown.

　　The mechanism has been confirmed, for a Ge nanowire grown by the VLS process, by tracking its evolution *in situ* within a TEM fitted with a high temperature stage (Wu and Yang 2001). The sequence of images in Figure 10.5 is in good agreement with the steps shown schematically in Figure 10.4.

10.2.2　THE TEMPLATE PROCESS

Formation of nanowires, and nanotubes, by the so-called template method is illustrated schematically in Figure 10.6. In essence, the wires are cast within confining channels; the template material is then removed subsequently, generally by chemical means. Various materials with porosity on the mesoscale have been used, including polycarbonate and alumina membranes (Martin et al. 1999), porous silica (Lee et al. 2001), cylindrical micelles (Kijima et al. 2004), and block copolymers (Cornelissen

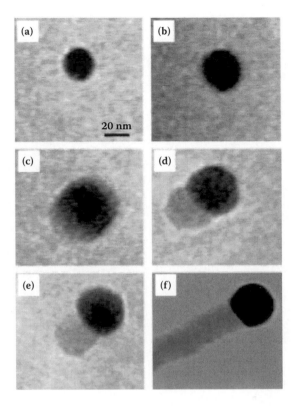

FIGURE 10.5 The nucleation and growth of a Ge nanowire by the VLS process, as shown in the sequence of TEM images (A) to (F). The gaseous precursor, GeI, was cracked catalytically at the Au seed particle, from which the wire was nucleated and grown at temperatures ranging from 970 to 1170 K. (From Wu and Yang, 2001, *J. Am. Chem. Soc.*, with permission from American Chemical Society.)

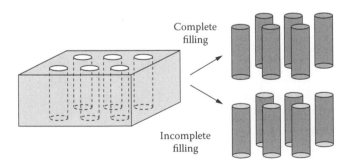

FIGURE 10.6 Schematic illustration of template formation of nanowires and nanotubes within nano/meso-scale porosity. (From Xia et al., 2003, *Adv. Materials*, with reference to Hulteen and Martin, 1997; and Cepak and Martin, 1999, with permission from John Wiley and Sons.)

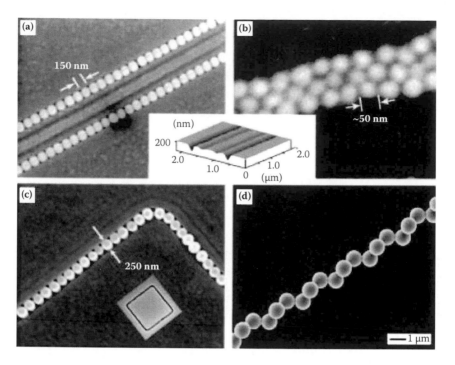

FIGURE 10.7   SEM images of 1-D structures self-assembled from nanoparticles. Chains of 150 nm polystyrene beads (A) and 50 nm Au colloidal particles (B) self-assembled along 120 nm wide lines patterned on a thin photoresist layer. The template is shown in the inset diagram. (C) A curved chain of Au-coated SiO$_2$ particles self-assembled along a square template groove (see inset), and (D) a spiral chain of self-assembled polystyrene particles in a V-groove template in a Si (100) substrate. (Adapted from Yin et al., 2001, *J. Am. Chem. Soc.*; Lu et al., 2002, *Nano Lett.,* Yin and Xia, 2003, *J. Am. Chem. Soc.*)

et al. 2002). The template method can also be exploited at linearly aligned preferred sites on a solid surface (e.g., a groove or a step) (Penner 2002).

A range of nanowires has been synthesized by the template method, ranging from metals (e.g., Gao et al. 2002), to semiconductors (e.g., Cao et al. 2001), ceramics (e.g., Zheng et al. 2001), and polymers (e.g., Sapp et al. 1999).

The feedstock can be delivered to the pores in a variety of ways, including vapour-phase sputtering, liquid-phase injection, or solution-phase chemical or electrochemical deposition. As well, a sol-gel route has been reported (Limmer et al. 2002). If the filling process is terminated before completion, then the outcome can be a tube structure, as shown in Figure 10.6.

The combination of template and self-assembly by nanoparticles is shown in Figure 10.7.

10.3   CHARACTERIZATION OF NANOWIRES BY SEM AND TEM

Nanowires tend to be relatively large objects in comparison with fullerene molecules, quantum dots, and carbon nanotubes. Thus, topographical and structural analyses of

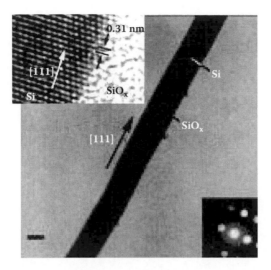

FIGURE 10.8 TEM characterization of nanowires synthesised by ablation of an $Si_{0.9}Fe_{0.1}$ target. The image shows diffraction contrast for a single wire; the crystalline Si core with the darker contrast is distinct from the brighter contrast of the amorphous SiO_2 sheath. The scale bar corresponds to 10 nm. The inset in the lower right-hand corner shows the diffraction pattern obtained along the [211] zone axis. The inset in the upper left-hand corner shows an HRTEM image, revealing a spacing of 0.31 nm in the direction perpendicular to the long axis of the wire. (Adapted from Morales and Lieber, 1998.)

wires by scanning electron microscopy (SEM) and transmission electron microscopy (TEM) are routinely relatively straightforward. For instance, selected area diffraction (SAD) can often be carried out. Some examples from the literature, shown in Figure 10.8 through Figure 10.11, illustrate the uses of electron-optical techniques.

10.4 CHARACTERIZATION OF NANOWIRE HETEROSTRUCTURES

Many of the recent developments in electronics and optoelectronics have been based on heterostructures. Thus, if nanowires are going to progress beyond a role as interconnects, and become devices in their own right, heterostructured wires will be a mandatory development. This has, in fact, been demonstrated (Wu et al. 2002), as shown in Figure 10.12.

10.5 CHARACTERIZATION RELATED TO POTENTIAL APPLICATIONS

Junctions between vertically oriented doped Si-nanowire arrays may have some potential as inexpensive and moderately efficient photovoltaic (PV) devices, due to the more forgiving requirements of purity and defect density of the Si feedstock. Alternatively, the configuration can be that of a radial heterostructure. The configurations and TEM characterization are shown in Figure 10.13.

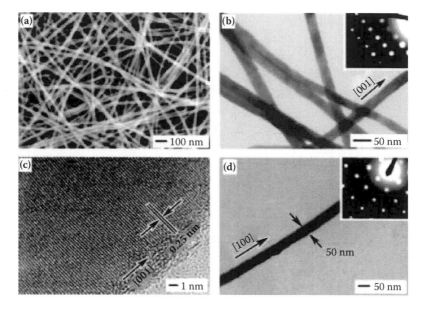

FIGURE 10.9   SEM and TEM characterization of α-Ag$_2$Se nanowires obtained by reacting trigonal t-Se phase nanowires with an aqueous AgNO$_3$ solution. (A) SEM overview image. (B) TEM image with diffraction pattern (inset). (C) HRTEM image obtain from the edge of a wire illustrating single crystal growth. The fringe spacing of 0.25 nm was consistent with the interplanar spacing of [200] planes, and a growth direction of <100> was apparent. (D) TEM image of a wire of greater diameter, ca. 50 nm, and a diffraction pattern (inset) showing tetragonal structure. (Adapted from Gates et al., 2001, *J. Am. Chem. Soc.*, Gates et al., 2002, *Adv. Funct. Mater.*, with permission from the American Chemical Society.)

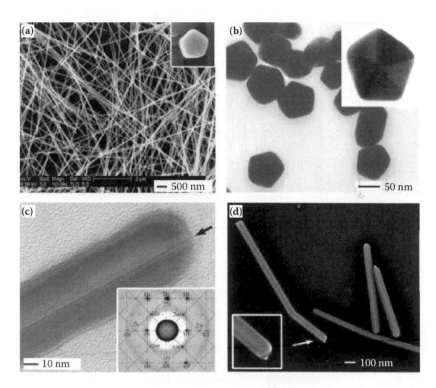

FIGURE 10.10 Analysis of Ag nanowires synthesised by the chemical passivation route (through exposure to poly(vinylpyrrolidene) (PVP), whereby growth is restricted to the end facets. Growth at the ends was promoted by selective attachment of Au catalyst particles (See Figure 10.1D). (A) SEM image demonstrating a uniform diameter (ca, 50 nm) of Ag. The inset shows a pentagonal cross-section. (B) TEM image of a microtomed thin slice of one of the wires in (A); the pentagonal symmetry of the cross-section is apparent. The inset shows a magnified TEM view, confirming that the wire has a fivefold twinned structure. (C) TEM image obtained from the end of a wire; the bright longitudinal line (identified by the arrow) is consistent with a twin plane being located along the long axis of the wire. The inset shows an SAD diffraction pattern recorded from a wire with the electron beam aligned perpendicular to one of the {100} facets. (D) SEM image of Ag nanowires after exposure to 1,12-dodecan-edithiol, in order to activate the Au nanoparticles in solution, followed by incubation with Au particles for 10 h. Attachment of Au particles was observed only at the ends of the wires, as bright dots (see inset), suggesting that the side faces were passivated by PVP coverage. (From Chen et al., 2007, *Langmuir*, with permission from the American Chemical Society.)

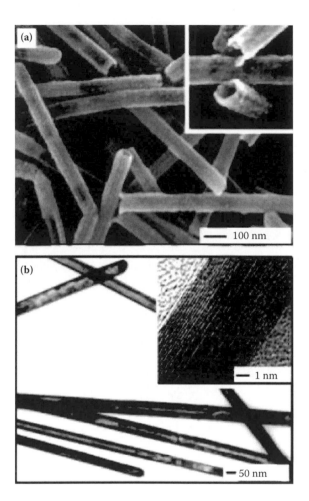

FIGURE 10.11 Crystalline metal nanotubes can be synthesised by a 'reverse' template method in which a metal (e.g., Ag) nanowire acts as the original template. An ion exchange process can be set up by dispersing the nanowires in an aqueous solution of, for instance, $HAuCl_4$ or $Pd(NO_3)_2$. The exposed silver atoms at the surface of the wire will then oxidise and ion exchange with Au ions (or Pd ions), and the latter will plate out on the Ag template. Nucleation takes place on the surface of the wire, and will, as the reaction progresses, form a cylindrical Au/Pd-sheath around the template (for additional description, see Sun et al., 2002). (A) SEM images of Pd nanotubes synthesised by a reverse template method. (B) TEM image of an Au nanotube. An HRTEM image (see inset) taken from an edge shows crystalline structure and uniformity of wall thickness. (From Sun et al., 2002, *Nano Lett.*, with permission from the American Chemical Society.)

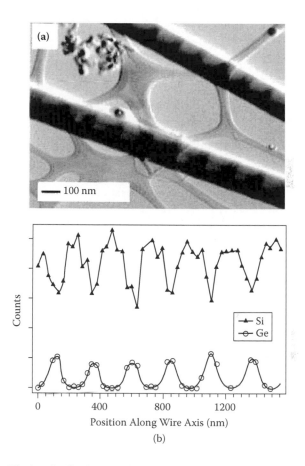

FIGURE 10.12 The longitudinal compositional modulation of an Si/SiGe superlattice structure synthesised by pulsed laser ablation in combination with CVD, in which the Si and Ge vapour sources were controlled independently, so that the requisite constituents could be delivered sequentially into a VLS growth environment. (A) Image from a TEM operated in the scanning transmission bright field mode shows two wires. The Ge-rich bands exhibit the darker contrast, due to Ge being a stronger scatterer. (B) Compositional EDS profiles in the longitudinal direction show the periodicity of the structure. (From Wu et al., 2002, *Nano Lett.*, with permission from the American Chemical Society.)

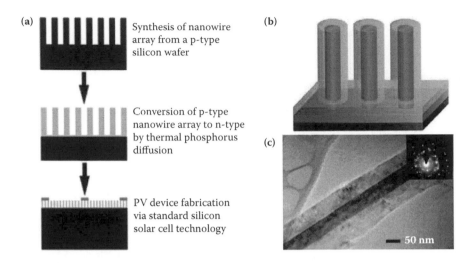

FIGURE 10.13 Aspects of solar cells based on nanowire arrays. (a) Schematic of fabrication process for a subsurface p-n junction device. (b) Schematic of a radialheterostructure nanowire array, and (c) TEM image of one of the synthesised nanowires showing the n-type crystalline core and p-type polycrystalline shell. (From Peng et al., 2005, *Small*, with permission from John Wiley and Sons.)

Nanowire arrays can also act as the element for direct light-to-energy conversion (see Kuykendall et al. 2007). A SEM image of an array of nanowires/rods is shown in Figure 10.14.

Lasing action has been observed in as-synthesised ZnO nanowires, and has demonstrated that a well-facetted wire can be an effective resonance cavity. Scanning near-field optical microscopy (SNOM) was used to demonstrate laser emission at the ends of a wire (see Figure 10.15), pumped by 0.5 ps pulses at a wavelength of 285 nm.

A recent review of the wider properties of semiconducting ZnO 1-D nanostructures illustrates the richness of the system (Wang 2009). As well as nanowires, structures such as nano-belts, -springs, and -helices, can be synthesised, and may lead to interesting and useful applications (e.g., chemical and biological nanosensors, solar cells, light emitting diodes, nanogenerators, and nanopiezotronic devices). Likewise, the Ag-based system can be synthesized as wires, belts, and beams (Chen et al. 2007). SEM and TEM analyses of Ag nanobeams are shown in Figure 10.16.

A great deal of activity is in progress to explore the potential of nanowires as elements of electronic devices (e.g., Hayden et al. 2008; Talin et al. 2010; Thelander et al. 2006; Lu and Lieber 2007). While it is not possible to cover all the methods that have been adopted to this end, it is instructive to consider a couple of examples.

Measurement of resistivity is an important aspect in determining the suitability of nanowires as interconnects. A typical experimental configuration is shown in Figure 10.17 together with some results.

An approximate expression for the resistance specific to a nanobeam can be derived (Lu and Lieber 2007), as follows:

A nanobeam will have length, L, width, W, and thickness, T (in the present case $W/T \approx 1.4$). The effective size-dependent resistivity is ρ, and the (lower) bulk resistivity

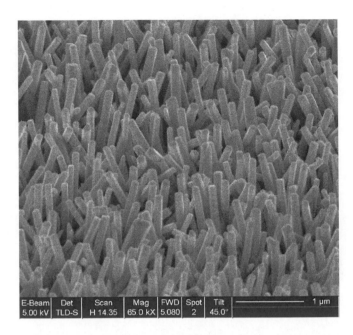

FIGURE 10.14   SEM image showing an InGaN nanowire/rod array. (From Kuykendall et al., 2007, *Nat. Mater.*, reprinted by permission from Macmillan Publishers Ltd.)

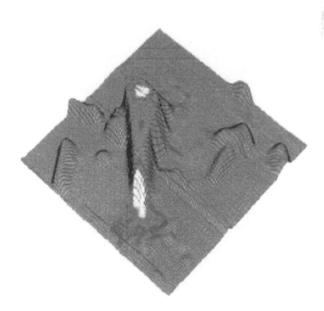

FIGURE 10.15   SNOM image of laser emission from the ends of a ZnO nanowire. (From Johnson et al., 2001, *J. Phys. Chem. B*, with permission from the American Chemical Society.)

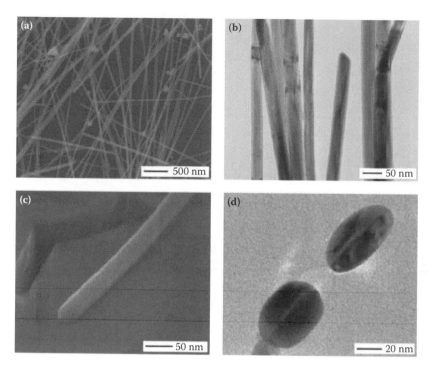

FIGURE 10.16 Analyses of Ag nanobeams by SEM (A) and TEM (B). A single beam is shown at a 65° tilt in the SEM image in (C). The cross-sectional profile of a microtome slice is shown in a TEM image in (D). (From Chen et al., 2007, *Langmuir*, with permission from the American Chemical Society.)

is ρ_0. For a given cross-sectional shape, there will be an equivalent cross-sectional area, $A = a^2$ of a square section beam. The correction to the bulk resistivity should initially be linear in the dependency of the ratio of surface area to volume (1/a for a cubic section). Thus, $\rho = \rho_0(1 + \lambda/a)$, where λ is a size-dependent scale factor, related to the bulk mean free path, ℓ, for carrier scattering. The scale factor will depend on ℓ, and also on the shape of the beam and surface scattering mechanism(s). In addition to the intrinsic resistance of the beam, there will be a contact resistance, R_0. Taking $A \approx \pi WT/4 = 0.56W^2$ and $a = A^{1/2} = 0.7W$, the following expression is obtained.

$$R - R_C \approx \rho \frac{L}{A} \approx \rho_o \frac{L}{0.56W^2}\left(1 + \frac{\lambda}{0.75W}\right) \tag{10.1}$$

Arguably, the most active area of device research has been concerned with field-effect transistor (FET) configurations and performance (e.g., Thelander et al. 2006). A radial configuration is shown in Figure 10.18.

A more conventional layout of a FET nanowire test device is shown in Figure 10.19.

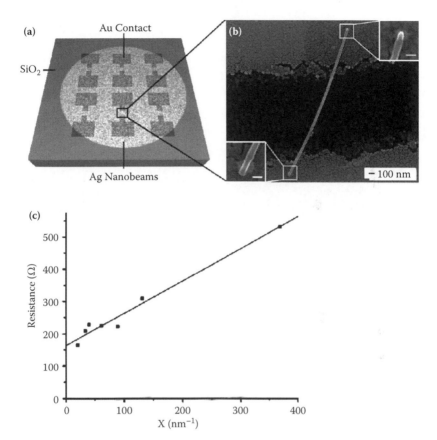

FIGURE 10.17 A typical arrangement for measurement of resistance. (A) Ag beam-shaped wires are deposited from solution onto gold contact pads. (B) The cross-sectional dimensions of a beam, e.g., 27×19 nm^2, are inferred from an SEM image. The higher resolution images in the insets, (50 nm scale bar), show deformation resulting from the fusing of the ends to the pads with a focused electron beam. (C) Plot of resistance versus $X = R - R_C$ is shown for a best fit with $\lambda = 15$ nm. The straight-line fit has a slope of unity, and the intercept gives a contact resistance of R_C of 164 Ω. (From Lu and Lieber, 2007, *Nat. Mater.*, reprinted by permission from Macmillan Publishers Ltd.)

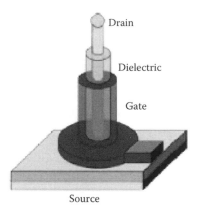

FIGURE 10.18 (Top) Schematic nanowire-based FET device with wrap-around gate dielectric. (Bottom) SEM image of vertical nanowire FET array with wrap-around gate. (Adapted from Bryllert et al., 2006, *Nanotechnology.*)

FIGURE 10.19 AFM image in a 3-D representation ($7 \times 7\ \mu m^2$ field of view) of a nanowire test device for exploration of FET configurations. An InP nanowire (of diameter 70 nm) is seen leading diagonally from left to right. Three metallic contact pads are used as ohmic current leads for two test structures on the same nanowire. The left hand structure has two closely spaced electrodes for four-point voltage measurement. The structure on the right has several side electrodes for local gating. (From Thelander et al., 2006, *Materials Today*, with permission from Elsevier.)

REFERENCES

Bryllert, T., Wernersson, L.-E., Löwgren, T. and Samuelson, L., 2006, *Nanotechnology*, 17, S227.

Cao, H., Xu, Y., Hong, J., Liu, H., Yin, G., Li, B., Tie, C. and Xu, Z., 2001, *Adv. Mater.*, 13, 1393.

Cepak, V. M. and Martin, C. R., 1999, *Chem. Mater.*, 11, 1363.

Chen, J., Wiley, B. J. and Xia, Y., 2007, *Langmuir*, 23, 4120.

Cornelissen, J. J. L. M., van Heerbeek, R., Kamer, P. C. J., Reek, J. N. H., Sommerdijk, N. A. M. and Nolte, R. J. M, 2002, *Adv. Mater.*, 14, 489.

Gao, T., Meng, G., Zhang, J., Sun, S. and Zhang, L., *Appl.Phys. A*, 2002, 74, 403.

Gates, B., Mayers, B., Wu, Y., Sun, Y., Cattle, B., Yang, P. and Xia, Y., 2002, *Adv. Funct. Mater.*, 12, 697.

Gates, B., Wu, Y., Yin, Y., Xia, Y, 2001, *J. Am. Chem. Soc.*, 123, 11500.

Hayden, O., Agarwal, R. and Luc, W., 2008, *Nano Today*, 3 (5–6), 12.

Hochbaum, A, I. and Yang, P., 2010, *Chem. Rev.,* 110, 527.

Hu, J., Odom, T. W. and Lieber, C. M., 1999, *Acc. Chem. Res.*, 32, 435.

Hulteen, J. C. and Martin, C. R., 1997, *J. Mater. Chem.*, 7, 1075.

Johnson, J. C., Yan, H., Schaller, D., Haber, L., Saykally, R. J. and Yang, P., 2001, *J. Phys. Chem. B*, 105, 11387.

Kijima, T., Yoshimura, T., Uota, M., Ikeda, T., Fujikawa, D., Mouri, S. and Uoyama, S., 2004, *Angew. Chem., Int. Ed.*, 43, 228.

Kuykendall, T., Ulrich, P., Aloni, S. and Yang, P. 2007, *Nat. Mater.,* 6, 951.

Lee, K.-B., Lee, S.-M. and Cheon, J., 2001, *Adv. Mater.*, 294, 348.

Limmer, S. J., Seraji, S., Wu, Y., Chou, T. P., Nguyen, C. and Cao, G., 2002, *Adv. Funct. Mater.*, 12, 59.

Lu, W. and Lieber, C. M., 2007, *Nat. Mater.,* 6, 841.

Lu, Y., Yin, Y., Li, Z.Y. and Xia, Y., 2002, *Nano Lett.,* 2, 785.

Martin, B. R., Dermody, C. J., Reiss, B. D., Fang, M., Lyon, L. A., Natan, M. J. and Mallouk, T. E., 1999, *Adv. Mater.*, 11, 1021.

Morales, A. M. and Lieber, C. M., 1998, *Science*, 279, 208.

Peng, K. Q., Xu, Y., Wu, Y., Yan, Y. J., Lee, S. T., and Zhu, J., 2005, *Small*, 1, 1062.

Penner, R. M. J., 2002, *Phys. Chem. B*, 106, 3339.

Sapp, S. A., Mitchell, D. T., and Martin, C. R., 1999, *Chem. Mater.*, 11, 1183.

Schmidt, V., Senz, S. and Gösele, U., 2005, *Z. Metallkd,,* 96, 427.

Schmidt, V., Wittemann, J. V., and Gösele, U., 2010, *Chem. Rev.*, 110, 361.

Sun, Y., Mayers, B. T., and Xia, Y, 2002, *Nano Lett.*, 2, 481.

Talin, A. A., Leonard, F., Katzenmeyer, A. M., Swartzentruber, B. S., Picraux, S. T., Toimil-Molares, M. E., Cederberg, J. G., Wang, X., Hersee, S. D. and Rishinaramangalum, A., 2010, *Semicond. Sci. Technol.*, 25, 024015.

Thelander, C., Agarwal, P., Brongersma, S., Eymery, J., Feiner, L. F., Forchel, A., Scheffler, M., Riess, W., Ohlsson, B. J., Gösele, U., and Samuelson, L., 2006, *Materials Today*, 9, 10, 28.

Wagner, R. S. and Ellis, W. C., 1964, *Appl. Phys. Lett.*, 4, 89.

Wang, Z. L., 2009, *Mater. Sci. Eng.*, 2009, R64, 33.

Wu, Y., Fang, R., and Yang, P., 2002, *Nano Lett.*, 2, 83.

Wu, Y. and Yang, P., 2001, *J. Am. Chem. Soc.*, 123, 3165.

Xia, Y., Yang, P., Sun, Y., Wu, Y., Mayers, F., Gates, B., Yin, Y., Kim, F., and Yan, H., 2003, *Adv. Materials*, 15, 353.

Yin, Y., Lu, Y., Gates, B., and Xia, Y., 2001, *J. Am. Chem. Soc.*, 123, 8718.

Yin, Y. and Xia, Y., 2003, *J. Am. Chem. Soc.,* 125, 2048.

Zheng, M., Zhang, L., Zhang, X., Zhang, J., and Li, G., 2001, *Chem. Phys. Lett.,* 334, 298.

11 Graphene and Other Monolayer Structures

11.1 INTRODUCTION

Graphene is the latest addition to the long list of carbon-based structures that have captured and held the attention of the research community (Wassei. and Kaner 2010; Wu, Yu, and Shen 2010), beginning with the many forms of diamond-like carbon (DLC), progressing through C_{60} and its derivatives, and ending, (at present), with carbon nanotubes and other tube structures.

The richness of carbon-based structures is due principally to the nature of the bonding and the consequential flexibility with which dimensionality and shape can be accommodated. Carbon in the ground state has four valence electrons, two in the 2s sub-shell and two in the 2p. Bonding with other carbon atoms takes place with one of the 2s electrons being promoted into an unoccupied 2p orbital, leading to sp hybrid orbitals. Depending on the number, 1 to 3, of p orbitals that are involved with the s orbital, there can be three configurations, sp, sp^2, and sp^3. The latter two configurations form three and four bonds with neighbouring carbon atoms, respectively, leading to graphene and diamond structures. The ideal graphene is a monatomic monolayer of carbon atoms with a simple hexagonal ('chicken-wire') lattice. However, ideal monolayer graphene crystals in the free state are, at best, metastable at ambient temperature (Marder 2000). Thus, graphene relaxes to more stable structures, such as graphite, fullerene, and nanotubes. The graphite structure consists of a stacking sequence of graphene layers held together by the van der Waals force. A single layer, or multiple layers, of graphene can roll up to form tubular structures, known as carbon nanotubes (CNTs). With the introduction of pentagons, a portion of a graphene sheet can be deformed into a fullerene molecule. All these structures are stable. Thus, graphene may be considered as the precursor to the more stable carbon phases.

Single graphene sheets were identified in 2004 (Novoselov et al. 2004; Novoselov et al. 2005; Zhang et al. 2005) by mechanical exfoliation from graphite when it had been stabilised on a crystalline substrate. Since then, a great deal of progress has been made in the development of alternative synthesis routes. In parallel, many of the exceptional properties of graphene have been explored, culminating in the award of the Nobel Prize in Physics for 2010.

11.2 GRAPHENE STRUCTURE

At first glance it might appear that the crystal and electronic structures of graphene should be identical, at best, or as a special case, at worst, to those for single wall CNTs, or *vice versa*. Indeed the unit cells are identical, see Figures 11.1 and

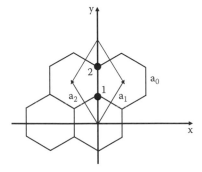

FIGURE 11.1 Structure of graphene. Atoms are located at the intersections, and sp² bonds connect the lattice points. The primitive unit cell is defined by vectors $\mathbf{a_1}$ and $\mathbf{a_2}$. There are two atoms per cell labelled 1 and 2. The nearest-neighbour distance a_0 is 0.142 nm.

Figure 9.4, as are the direct and reciprocal primitive unit cell vectors (shown below). However, the translational symmetries are very different. In the case of the CNT the unit cells are effectively tiles on a 2-D closed cylindrical surface, and the tight binding approximation provides considerable insight into the electronic structure. On the other hand, the same unit cell tiles describe the graphene structure on a flat 2-D plane of infinite extent. It turns out in the latter case that a rather more sophisticated Dirac formalism is required in order to describe the electronic structure. In the interest of presenting a self-contained discussion in this chapter, some repetition of material in Chapter 9 will appear.

$$b_1 = \left(\frac{1}{\sqrt{3}}, 1 \right) \frac{2\pi}{a} \tag{11.1}$$

$$b_2 = \left(\frac{1}{\sqrt{3}}, -1 \right) \frac{2\pi}{a} \tag{11.2}$$

where $a = \sqrt{3}a_0$, a_0 being the nearest-neighbour distance. The first Brillouin zone is hexagonal with a side length of $4\pi/3a$. There are two points, defined by the primitive reciprocal vectors,

$$K = \frac{2\pi}{3} \left(\frac{1}{\sqrt{3}a_0}, \frac{1}{3a_0} \right) \tag{11.3}$$

and

$$K' = \frac{2\pi}{3} \left(\frac{1}{\sqrt{3}a_0}, -\frac{1}{3a_0} \right) \tag{11.4}$$

within the first BZ where the A and B lattices decouple. These define the so-called Dirac points.

11.3 SUMMARY OF ELECTRONIC STRUCTURE

The band structure of graphene has already been derived in the tight-binding approximation (see Chapter 9). Linear combinations of Bloch wave functions are constructed for the $2p_z$ orbitals at the 1 and 2 sites in the primitive unit cell, and the Hamiltonian is limited to nearest-neighbour interactions. Solutions to the eigenvalue problem describe the π^* and π bands, which account for the in-plane electrical conductivity of graphene (Wallace 1947; Slonczewski and Weiss 1958).

The electronic dispersion in 3-D representation is shown on the left in Figure 11.2c, with the energy contour lines in momentum space to the right. Near the K and K' points, the energy dispersion has a conical shape. To a first-order approximation, the dispersion is given by

$$E(k) = \pm v_F |k| \tag{11.5}$$

where

$$v_F = \frac{3E_{c-c}a}{2h} \approx 10^6 \text{ m/s} \tag{11.6}$$

The wave vector k is measured from the K and K' points, and E_{c-c} is the C-C bond energy. The linear k-dependence of the energy dispersion for graphene is different from that of non-relativistic electrons, (i.e., $E(k) = h^2k^2/2m$), as shown in Figure 11.2f. Further away from the K and K' points, second-order contributions to the Hamiltonian give rise to some distortion of the dispersion relation.

The full flavour of the electronic structure and dynamics of graphene requires a rather more sophisticated derivation using the Dirac equation formalism and the treatment of the free electrons as relativistic Weyl fermions. (see Wu, Yu, and Shen 2010 and references therein).

11.4 OTHER 2-D STRUCTURES (NANOSHEETS)

As in the case of CNTs, where their synthesis and description have led rapidly to the discovery of a host of other tube structures (see Chapter 9), so it has been for the synthesis and description of graphene.

Layer structures in which sheets are stacked under the influence of van der Waals interaction, for example, BN and the transition metal chalcogenides, are prime candidates for delamination, and have indeed been produced in their single atomic layer forms. But delamination has also been reported for compounds that can be regarded as having layer structures, such as smectite clay minerals, metal phosphates, and

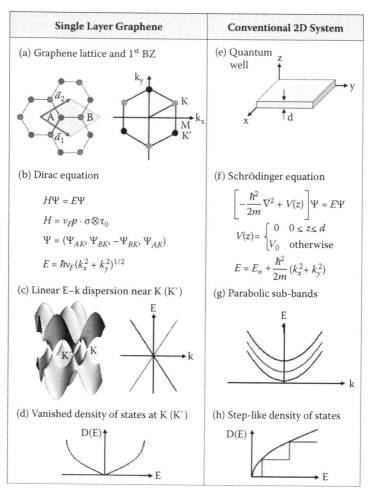

Single Layer Graphene	Conventional 2D System
(a) Graphene lattice and 1st BZ	(e) Quantum well
(b) Dirac equation $$H\Psi = E\Psi$$ $$H = v_F p \cdot \sigma \otimes \tau_0$$ $$\Psi = (\Psi_{AK}, \Psi_{BK}, -\Psi_{BK}, \Psi_{AK})$$ $$E = \hbar v_F (k_x^2 + k_y^2)^{1/2}$$	(f) Schrödinger equation $$\left[-\frac{\hbar^2}{2m}\nabla^2 + V(z) \right]\Psi = E\Psi$$ $$V(z) = \begin{cases} 0 & 0 \le z \le d \\ V_0 & \text{otherwise} \end{cases}$$ $$E = E_n + \frac{\hbar^2}{2m}(k_x^2 + k_y^2)$$
(c) Linear E–k dispersion near K (K`)	(g) Parabolic sub-bands
(d) Vanished density of states at K (K`)	(h) Step-like density of states

FIGURE 11.2 Comparison of graphene (a)–(d), and conventional 2-D (e)–(h), electron systems. (a) Lattice structure and first BZ; (b) Dirac equations; (c) 3-D (left) and 2-D (right) energy dispersions (note the K' points where the density of states vanishes); (d) DOS as a function of energy; (e) Schematic of a conventional 2-D quantum well confined by electrostatic potentials in the z direction; (f) Schrödinger equation, the potential function and the eigenvalue expression; (g) Energy as a function of **k**-vector; (h) DOS. (From Wu et al., 2010, *J. Appl. Phys.*, with permission from the American Institute of Physics.)

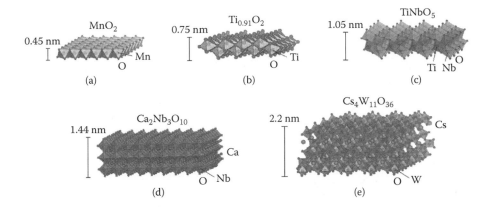

FIGURE 11.3 **(See color insert.)** Examples of oxide nanosheets. (a) MnO_2, (b) $Ti_{1-\delta}O_2$ (δ denotes Ti deficiency), (c) $TiNbO_5$, (d) $Ca_2Nb_3O_{10}$, (e) $Cs_4W_{11}O_{36}$. (From Osada and Sasaki, 2009, *J. Mater. Chem.*, reproduced with permission from the Royal Society of Chemistry.)

oxides (see review by Osada and Sasaki 2009, and references therein). Examples of oxide layer structures are shown in Figure 11.3.

11.5 OVERVIEW OF SYNTHESIS ROUTES

As for CNTs and other 1-D and 0-D structures, the original discovery of graphene, produced somewhat serendipitously, has been followed by a proliferation of other synthesis routes, and graphene can now be obtained by several different such routes. It would be fair to say that methods for producing graphene represent work in progress. Accordingly, this section cannot give justice to the topic. However, the present overview does serve as a vehicle for demonstrating the need for materials characterization of products that is emerging from present and future routes.

11.5.1 Mechanical Exfoliation

Application of the original method of mechanical exfoliation from highly oriented pyrolytic graphite generally yields small amounts of high-quality graphene flakes. In its simplest form, the method consists of bulk delamination of graphite by rubbing against a substrate. Another variation is to use adhesive tape to separate the layers, which can produce delamination at the monolayer level.

11.5.2 Liquid Phase Exfoliation

Liquid exfoliation and reduction of graphene oxide have been used to produce chemically converted graphene in quantities that have the potential for scaling up for industrial use (Stankovich et al. 2007). Flakes of BN monolayers have been prepared by liquid phase exfoliation (Hernandez et al. 2008) of bulk hexagonal BN in N-methyl-2-pyrrolidone.

11.5.3 Epitaxial Growth

The growth of the graphene structure is favoured on those substrates that have a reasonable lattice matching, as has been demonstrated, for instance, for graphene grown on the Ni(111) or Ir(111) surfaces (Grüneis and Vyalikh 2008) and on the hexagonal 6H- and 4C-SiC(0001) (α-SiC) surfaces (Rollings et al. 2006; Hass et al. 2006; Virojanadara et al. 2008; Emtsev et al. 2009). Formation of graphene layers by the thermal decomposition of α-SiC has been proposed for the synthesis of high-quality wafer-size layers. A particular merit of the method is that α-SiC is a large-gap semiconductor, whose property effectively decouples the electronic structure of the graphene layer from that of the substrate.

More recently, it has been shown that graphene can also be grown on cubic 3C-SiC (β-SiC) in spite of the apparent lack of lattice matching (Aristov et al. 2010). It was found that the interaction with the substrate was nearly negligible, thus permitting transfer of the graphene to other substrates. As well, nucleation and growth of graphene on Cu (111) has been reported (Gao, Guest, and Guisinger 2010).

11.5.4 Nucleation and Growth of Graphene on SiC

A single crystal SiC face can be terminated by either Si or C. Heat treatment at a temperature in the range 1000–1500°C in UHV, or in a reducing atmosphere, will lead to decomposition, and the initial formation of an amorphous layer of carbon on a Si substrate. The substrate then acts as a template for the epitaxial nucleation and growth of a graphene layer, via a number of intermediate stages; sometimes the presence of gaseous disilane is required. Further details and discussion of the synthesis route can be found in the literature (de Heer et al. 2007; Hass, de Heer, and Conrad 2008).

11.5.5 Catalyst Promoted Nucleation and Growth of Graphene

The predominant mechanism for chemical vapour deposition (CVD) growth on transition metals in the presence of an organic precursor is due to the solubility of carbon in the substrate. When the sample is cooled the carbon segregates to the surface, leading to the nucleation of graphene. The source of carbon arises from the dissociation of organic precursor molecules on the substrate.

By introducing Ni and Cu as the substrates for CVD growth, the size, thickness, and quality of the graphene so produced is beginning to approach industrially useful specifications (Reina et al. 2009; Li et al. 2009; Kim et al. 2009). Monolayer graphene can be grown on a substrate from a solid carbon source in the presence of a metal catalyst, for example, as shown in Figure 11.4 (Sun et al. 2010). The source

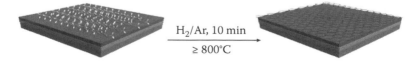

H_2/Ar, 10 min
———————————
≥ 800°C

FIGURE 11.4 (**See color insert.**) Schematic of the process route for catalyst-promoted, substrate-mediated synthesis of graphene. (Adapted from Sun et al., 2010.)

was a poly(methyl methacrylate) (PMMA) thin film (ca. 100 nm) spin-coated onto a thin Cu catalyst film. Heating at a temperature in the range 800 to 1000°C for 10 min in a low-pressure reducing gas flow (H_2/Ar), resulted in the formation of a uniform layer of graphene on the substrate. The graphene could subsequently be transferred to another substrate for characterization.

11.6 STRUCTURAL CHARACTERIZATION

The latest generation aberration-corrected and energy-filtered (scanning) transmission electron microscopy ((S)TEM) instruments have arguably been the most important tools for the structural characterization of monolayer nanosheets. Under favourable conditions, single-atom resolution can be obtained, and dynamic processes can be tracked at a temporal resolution in the one-second range. Examples for graphene are shown in Figure 11.5 through Figure 11.7.

The outcome of an ADF STEM analysis of a BN monolayer sample is shown in Figure 11.8 (Krivanek et al. 2010). The novel feature of this work is that single atom compositional information can be obtained by mapping the scattering intensity, rather than by electron energy loss spectroscopy (EELS). The intensity depends on the atomic number, (with a dependence of ca. $Z^{1.7}$), as is evident from the line trace in Figure 11.8c. Accordingly, it is possible to assign atomic numbers to single atom impurities, even for low-Z species such as C and O.

Scanning probe microscopy (SPM) in general, and atomic force microscopy (AFM) in particular, are widely used to map surface topography of nanostructures in 3-D. While the latest generation of SEM instruments can provide lateral resolution of ca. 1 nm under favourable conditions, AFM has the advantage of being able to resolve z-direction heights to better than 0.1 nm, while its depth of focus is equal to that of the z-piezo dynamic range (typically 5–10 μm). Another advantage is that analysis can be performed under ambient laboratory conditions of specimens that

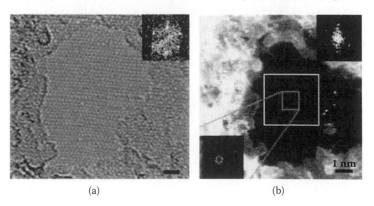

(a)   (b)

FIGURE 11.5 HRTEM imaging of monolayer graphene. (Left) Bright-field conditions. (Right) High-angle annular dark-field (HAADF) conditions. The insets in the top right-hand corners reveal the expected hexagonal symmetry, while the inset in the lower left-hand corner of b shows the structure at high resolution in direct space. (Adapted from Gass et al., 2008, *Nature Nanotech.*, reprinted by permission from Macmillan Publishers Ltd.)

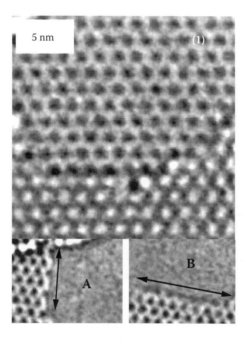

FIGURE 11.6 Aberration-corrected HRTEM images of graphene. Individual carbon atoms are resolved as white spots. Image (1) shows an overview of a single layer in the upper field of view, and a bilayer in the lower field of view. Images (A) and (B) illustrate armchair and zig-zag termination at edges of a hole in a flake. The zigzag edge, roughly 7 hexagons long, makes a 60° turn at the lower right-hand corner. (Adapted from Girit et al., 2009, and Meyer et al., 2008.)

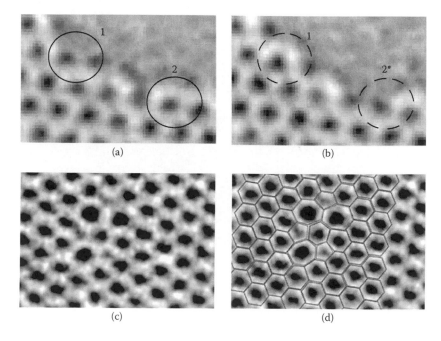

FIGURE 11.7 Successive images in (a) and (b) correspond to two frames in a sequence of the evolution of an edge structure. Two carbon atoms are removed from a location, solid circle 2, and are transferred to a location within solid circle 1, thus completing a hexagon. The net result is shown in (b), within the broken circles 1' and 2'. The images in (c) and (d) illustrate a transient defect structure that is associated with pentagons and heptagons. (Adapted from Girit et al., 2009, and Meyer et al., 2008.)

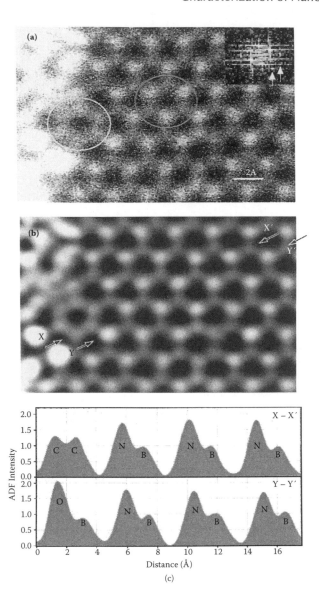

FIGURE 11.8 ADF STEM image of monolayer BN. (a) As recorded. (b) Corrected for distortion, smoothed, and deconvoluted to remove probe tail contributions to nearest neighbours. (c) Line profiles showing the image intensity (normalised to unity for a single boron atom) as a function of position in image (b) along X–X' and Y–Y'. The elements giving rise to the peaks seen in the profiles are identified by their chemical symbols. Inset at top right in (a) shows the Fourier transform of an image area away from the thicker regions. The two arrows point to the $(11\overline{2}0)$ and $(20\overline{2}0)$ reflections of the hexagonal BN structure, that correspond to recorded spacings of 0.126 and 0.109 nm, respectively. The length of the scale bar in (a) corresponds to 0.2 nm. The image was recorded at 60 keV beam energy, which is below the knock-on radiation damage threshold for BN, and the probe diameter was ca. 0.12 nm. (From Krivanek et al., 2010, *Ultramicroscopy*, reprinted by permission from Macmillan Publishers Ltd.)

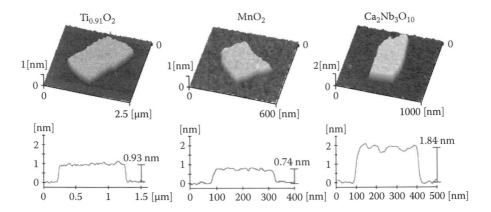

FIGURE 11.9 AFM images of $Ti_{0.91}O_2$, MnO_2, and $Ca_2Nb_3O_{10}$ nanosheets. The imaging was carried out in the tapping mode under vacuum conditions. Height profiles are shown in the bottom panels, and were consistent with known layer spacings. (From Osada and Sasaki, 2009, *J. Mater. Chem.*, with permission from the Royal Society of Chemistry.)

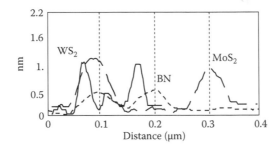

FIGURE 11.10 Monolayer heights of small exfoliated flakes revealed by AFM contour lines taken from images obtained in the tapping mode. The heights are consistent with known c-axis spacings. (From Myhra, S., 2010, unpublished results.)

present difficulties under e-beam conditions (i.e., insulators and organic/biomaterials). Many nanostructures are relatively non-reactive in air, and are therefore ideally suited for AFM analysis. See Chapter 5 for additional details.

Representative examples are shown in Figures 11.9 and 11.10, for some oxide nanosheets and for two metal chalcogenides and BN flakes, both exfoliated by wet chemistry and deposited on a Si wafer substrate.

11.7 RAMAN SPECTROSCOPIC CHARACTERIZATION

Raman spectroscopy is a user-friendly and non-destructive method for the characterization of carbon-based structures (see Ferrari and Robertson 2004). Raman spectra of carbon-based materials show characteristic features in the 800–2000 cm^{-1} region, designated as the G and D peaks, which are found at ca. 1560 and 1360 cm^{-1}, respectively. The G peak corresponds to the E_{2g} phonon at the centre of the Brillouin zone. The D peak arises from the breathing modes of the sp^2 hexagons, but

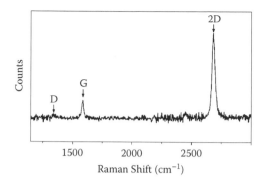

FIGURE 11.11 Raman signature of graphene from the catalyst-promoted, substrate-mediated synthesis route. The two most prominent peaks in this spectrum are the G peak at 1,580 cm^{-1} and the 2D peak at 2,690 cm^{-1}. The 2D/G intensity ratio was ca. 4 and the full-width at half-maximum of the 2D peak was ca. 30 cm^{-1}, consistent with the graphene being a monolayer. The intensity of the D peak (at 1,350 cm^{-1}) was only just distinguishable from the background, indicating the presence of sp^3 carbon atoms or defects at only trace levels. (adapted from Sun et al., 2010, *Nature*, with permission from American Chemical Society).

requires a defect to be active (Ferrari and Robertson 2000, 2001). Accordingly, the magnitude of the D mode is a measure of the quality of the graphene, in the sense that an absence, or a low intensity, of the D mode is an indicator of the absence of defects (Figure 11.11). The most prominent feature for graphene is the second-order 2D peak, located at ca. 2700 cm^{-1}. It is always present, and is a useful diagnostic for discriminating single versus multiple layers, as shown by the trends in Figure 11.12 which shows the spectra for increasing number of graphene layers in the specimen. The relative intensity of the 2D band decreases, while its width increases, with increasing numbers of layers.

It has been shown (Ni et al. 2007) that the Raman signature of graphene can be used to determine the number of layers, as illustrated in Figure 11.12. In essence, the intensity of the G-band is roughly proportional to the number of layers, at least it is so up to four.

11.8   CHARACTERIZATION OF ELECTRONIC STRUCTURE

The unusual electronic structure of graphene gives rise to a number of exceptional physical properties and phenomena that distinguish it from conventional 2-D electron gas systems, such as quantum dots (see Chapter 8). Some of the unique physical phenomena that have been observed or explored so far include unconventional integer quantum Hall effect (Novoselov et al. 2004; Zhang et al. 2005), Klein tunnelling (Katsnelson, Novoselov, and Geim 2006; Beenakker 2006; Stander, Huard, and Goldhaber-Gordon 2009), valley polarisation (Rycerz, Tworzydlo, and Beenakker 2006; Cresti, Grosso, and Parravicini 2008), universal (non-universal) minimum conductivity (Geim and Novoselov 2007), weak (weak anti-) localisation (Geim and Novoselov 2007; Suzuura and Ando 2002), ultrahigh mobility (Bolotin et al.

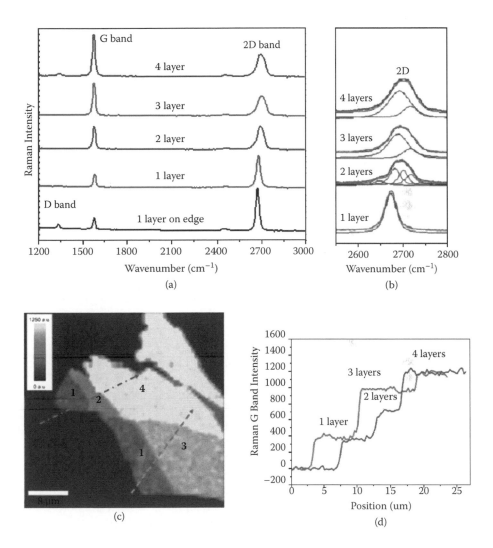

FIGURE 11.12 (**See color insert.**) (a). Raman spectra as a function of number of layers. (b) Details of the structure of the 2D band. The added complexity of the 2D structure is due to intralayer interactions, which perturb the in-layer modes. (c) Raman image generated by the intensity of the G band. (d) Cross sections of Raman image along the broken lines in (c). (From Ni et al., 2007, *Nano Lett.*, with permission from the American Chemical Society.)

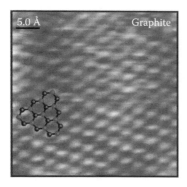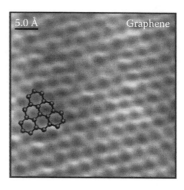

FIGURE 11.13 Comparison of the STM images from graphite (two or more graphene layers) and a single graphene layer on a non-interacting substrate. (a) Graphite structure: the two non-equivalent atom sites are shown. (b) Graphene structure: all the carbon atom sites are equivalent. (From Aristov et al., 2010, *Nano Lett.*, with permission from the American Chemical Society.)

2008; Orlita et al. 2008; Du et al. 2008), and specular Andreev reflection at the graphene–superconductor interface (Beenakker 2006; Zhang et al. 2008). Progress in the understanding of these effects has been reviewed by Katsnelson (2007).

The kinds of measurement and the methods required to explore the effects listed in the above paragraph are beyond the scope of this monograph, and the reader is referred to the literature. The discussion of the characterization of electronic structure will be confined here to the basic measurements that can be carried out in order to validate the outcomes of various synthesis routes.

UHV-STM cannot be considered to be user-friendly as a routine technique. However, graphene is relatively inert and can generally be analysed under ambient conditions by STM. Structural imaging by STM of single layer graphene and multilayer graphite is illustrated in Figure 11.13.

Conventional X-ray photoelectron spectroscopy (XPS) is a user-friendly and widely available technique. XPS is generally described as a surface-specific technique with a depth of information volume of a few monolayers. However, in the context of graphene and other mono/few-layer structures, it must be considered to be a bulk technique. The principal merit of XPS is that it is a quantitative compositional probe for all elements, except H and He. Moreover, it provides chemical information that can be derived from the chemical shifts of elemental binding energies, allowing determination of the chemical electronic environments of the species present in the sample (see, for instance, Briggs and Grant 2003; Rivière and Myhra 2009).

Conventional XPS is normally regarded as a broad-beam technique with a lateral resolution in the few-μm range. Thus, it is rarely relevant for analysis of single monolayer flakes (but can be quite useful for extended specimens consisting of a number of flakes).

Synchrotron-based spectroscopies have a number of advantages such as added luminosity leading to improved spectral resolution at higher count rates, a wide and continuously variable excitation beam energy, greater count rates and better signal-to-noise ratios, and a generally smaller spot size for the incident beam. The downside is that 'free' access to synchrotron sources is limited (full-cost-recovery access is generally prohibitively expensive), and there is usually a lengthy set-up procedure. Thus, routine measurements at a synchrotron source are rarely an option, although it is an invaluable means for carrying out high-priority fundamental science. The spectra shown in Figure 11.14 illustrate the chemical information that can be obtained with synchrotron spectroscopy (c) and (d) and, to a lesser extent, with conventional XPS (a) and (b).

The system that generated the spectra in Figure 11.14 consisted of a single crystal SiC substrate on which a monolayer was being nucleated and grown. The spec-

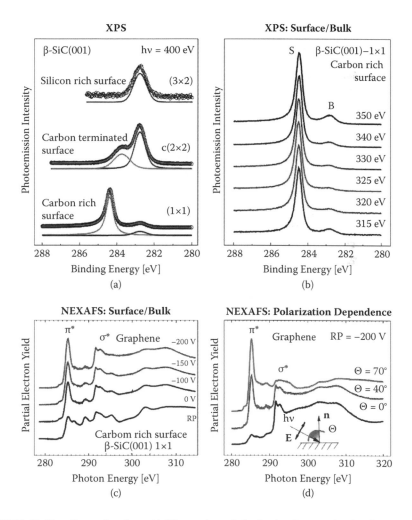

FIGURE 11.14 (**See color insert.**) Illustration of the basic characterization by synchrotron photoelectron spectroscopy of the electronic structure of graphene synthesised on β-SiC substrate. (From Aristov et al., 2010, *Nano Lett.*, with permission from the American Chemical Society.)

tral finger-prints provided information about the state of the substrate and about the evolution of graphene synthesis during the process.

In Figure 11.14, the C 1s photoemission spectra shown in (a) were recorded at an incident photon energy of 400 eV, from the β-SiC (001) substrate at different stages of the synthesis process. The data show an evolution from a carbide surface (low C 1s binding energy) via the appearance of a carbon surface layer (shoulder at ca. 284 eV) to a predominantly graphitic signal at ca. 284.6 eV. These assignments are compatible with results from standard spectra in the literature (e.g., Wagner et al. 1979). The results in (b) of the evolution of the C 1s photoemission structure of the C-rich surface as a function of photon energy essentially confirm the results in (a). As the incident photon energy is tuned above and below its value, 325eV, corresponding to maximum surface sensitivity, (i.e., minimum mean inelastic escape depth), the carbide contribution, B, increases due to more of the substrate carbon becoming included in the interaction volume. The binding energy of the more surface-specific component, S, is consistent with that for graphene/graphite. The NEXAFS C 1s spectra shown in (c) were recorded for increasing retard potential, (RP), in order to separate bulk and surface features. For the most surface sensitive potential, (at RP = −200 eV), the spectral signature was consistent with that for graphene/graphite. The angle-resolved spectra shown in (d) reveal a pronounced anisotropy for the linearly polarized incident radiation. The angular dependence of the resonances suggests that the π* orbitals are oriented perpendicular (i.e., out-of-plane), and the σ* orbitals are oriented parallel (i.e., in-plane), to the surface, as expected for graphene/graphite (Aristov et al. 2010).

REFERENCES

Aristov, V. Y., Urbanik, G., Kummer, K., Vyalikh, D. V., Molodtsova, O. V., Preobrajenski, A. B., Zakharov, A. A., Hess, C., Hänke, T., Büchner, B., Vobornik, B. I., Fujii, J., Panaccione, G., Ossipyan, Y. A., and Knupfer, M., 2010, *Nano Lett.*, 10, 992.

Beenakker, C. W. J., 2006, *Phys. Rev. Lett.*, 97, 067007.

———, 2008, *Rev. Mod. Phys.*, 80, 1337.

Bolotin, K. I., Sikes, K. J., Jiang, Z., Klima, M., Fudenberg, G., Hone, J., Kim, P., and Stormer, H. L., 2008, *Solid State Commun.*, 146, 351.

Briggs, D. and Grant, J. T., (eds.), 2003, *Surface Analysis by Auger and X-Ray Photoelectron Spectroscopy*, Chicester: IM Publishing.

Cresti, A., Grosso, G., and Parravicini, G. P., 2008, *Phys. Rev. B,.* 77, 233402.

de Heer, W. A., Berger, C., Wu, X. S., First, P. N., Conrad, E. H., Li, X. B., Li, T. B., Sprinkle, M., Hass, J., Sadowski, M. L., Potemski, M., and Martinez, G., 2007, *Solid State Commun.*, 143, 92.

Du, X., Skachko, I., Barker, A., and Andrei, E. Y., 2008, *Nature Nanotechnol.*, 3, 491.

Emtsev, K. V., Bostwick, A., Horn, K., Jobst, J., Kellogg, G. L., Ley, L., McChesney, J. L., Ohta, T., Reshanov, S. A., Röhrl, J., Rotenberg, E., Schmid, A. K., Waldmann, D., Weber, H. B., and Seyller, T., 2009, *Nature Mater.*, 8, 203.

Ferrari, A. C. and Robertson, J., 2000, *Phys. Rev. B*, 61, 14095.

———, 2001, *Phys. Rev. B*, 64, 075414.

———, (eds.), 2004, Special issue in *Phil. Trans. R. Soc. London, Ser. A*, 362, 2267.

Gao, L., Guest, J. R., and Guisinger, N. P., 2010. *Nano Lett.*, 10, 3512.

Gass, M. H., Bangert, U., Bleloch, A. L., Wang, P., Nair, R. R., and Geim, A. K., 2008, *Nature Nanotech.*, 3, 676.

Geim, A. K. and Novoselov, K. S., 2007. *Nature Mater*. 6, 183.

Girit, C. O, Meyer, J. C., Erni, R., Rossell, M. D., Kisielowski, C., Yang, L., Park, C. H., Crommie, M. F., Cohen, M. L., Louie, S. G., and Zettl, A, 2009, *Science*, 323, 1705.

Grüneis, A. and Vyalikh, D. V., 2008, *Phys. Rev. B*, 77, 193401.

Hass, J. Feng, R., Li, T., Li, X., Zong, Z.. de Heer, W. A., First, P. N., Conrad, E. H., Jeffrey, C. A., and Berger, C., 2006, *Appl. Phys. Lett.*, 89, 143106.

Hass, J., de Heer, W. A., and Conrad, E. H., 2008. *J. Phys. Condens. Matter*, 20, 323202.

Hernandez, Y. Nicolisi, V., Lotya, M., Blighe, F. M., Sun, Z., De, S., McGovern, I.T., Holland, B., Byrne, M., Gunko, Y. K., Boland, J. J., Niraj, P., Duesberg, G., Krishnamurthy, S., Goodhue, R., Hutchison, J., Scardaci, V., Ferrari, A. C., and Coleman, J. N. 2008. *Nature Nanotechnol.*, 3, 563.

Katsnelson, M. I., 2007, *Materials Today*, 10, 20.

Katsnelson, M. I., Novoselov, K. S., and Geim, A. K., 2006, *Nature Phys.* 2, 620.

Kim, K. S., Zhao, Y., Jang, H., Lee, S. Y., Kim, J. M., Kim, K. S., Ahn, J. H., Kim, P., Choi, J. Y., and Hong, B. H., 2009, *Nature,* 457, 706.

Krivanek, O. L., Dellby, N., Murfitt, M. F., and Chisholm, M. F., 2010, *Ultramicroscopy*, 110, 935.

Li, X. Cai, W., An, J., Kim, S., Nah, J., Yang, D., Piner, R., Velamakanni, A., Jung, I., Tutuc, E., Banerjee, S. K., Colombo, L., and. Ruoff, R. S., 2009, *Science* 324, 1312.

Marder, M. P., 2000, *Condensed Matter Physics*, New York: Wiley.

Meyer, J. C., Kisielowski, C., Erni, R., Rossell, M. D., Crommie, M. F., and Zettl, A., 2008, *Nano Lett.*, 8, 3582.

Ni, Z. H.. Wang, H. M., Kasim, J., Fan, H. M., Yu, T., Wu, Y. H., Feng, P., and Shen, Z. X., 2007, *Nano Lett.*, 7, 2758.

Novoselov, K. S., Geim, A. K., Morozov, S. V., Jiang, D., Zhang, Y., Dubonos, S. V., Grigorieva, I. V., and Firsov, A. A., 2004, *Science* 306, 666.

Novoselov, K. S., Geim, A. K., Morozov, S. V., Jiang, D., Katsnelson, M. I, Grigorieva, I. V., Dubonos, S. V., and Firsov, A. A., 2005, *Nature* (London), 438, 197.

Orlita, M., Faugeras, C., Plochocka, P., Neugebauer, P., Martinez, G., Maude, M., Barra, A. L. Sprinkle, M., Berger, C., de Heer, W. A., and Potemski, M., 2008, *Phys. Rev. Lett.*, 101, 267601.

Osada, M. and Sasaki, T., 2009, *J. Mater. Chem.*, 19, 2503.

Reina, A., Jia, X., Ho, J., Nezich, D., Son, H., Bulovic, V., Dresselhaus, M. S., and Kong, J., 2009, *Nano Lett.*, 9, 30.

Rivière, J. C. and Myhra, S., (eds.), 2009, *Handbook of Surface and Interface Analysis*, 2nd ed., Boca Raton, FL: CRC Press.

Rollings, E., Gweon, G. H., Zhou, S. Y., Mun, B. S., McChesney, J. L., Hussain, B. S., Fedorov, A. V., First, P. N., de Heer, W. A., and Lanzara, A., 2006, *J. Phys. Chem. Solids*, 67, 2172.

Rycerz, A., Tworzydlo, J., and Beenakker, C. W. J., 2007, *Nature Phys.*, 3, 172.

Slonczewski, J. C. and Weiss, P. R., 1958, *Phys. Rev.*, 109, 272.

Stander, N., Huard, B., and Goldhaber-Gordon, D., 2009, *Phys. Rev. Lett.*, 102, 026807.

Stankovich, S. Dikin, D. A., Piner, R. D., Kohlhaas, K. A., Kleinhammes, A., and Ruoff, R. S., 2007, *Carbon,* 45, 1558.

Sun, Z., Yan, Z., Yao, J., Beitler, E., Zhu, Y., and Tour, J. M., 2010, *Nature*, 468, 549.

Suzuura, H. and Ando, T., 2002, *Phys. Rev. Lett.*, 89, 266603.

Virojanadara, C., Syväjarvi, M., Yakimova, R., Johansson, L. I., Zakharov, A. A., and Balasubramanian, T., 2008, *Phys. Rev. B*, 78, 245403.

Wagner, C. D., Riggs, W. M., Davis, L. E., Moulder, J. F., and Muilenberg, G. E., 1979, *Handbook of X-Ray Photoelectron Spectroscopy*, Eden Prairie, U.S.: Perkin-Elmer Corporation.

Wallace, P. R., 1947, *Phys. Rev.*, 71, 622.

Wassei, J. K. and Kaner, R. B., 2010, *Materials Today*, 13, 52.

Wu, Y. H., Yu, T., and Shen, Z. X., 2010, *J. Appl. Phys.*, 108, 071301.

Zhang, Q. Y., Fu, D. Y., Wang, B. G., Zhang, R., and Xing, D. Y., 2008, *Phys. Rev. Lett.*, 101, 047005.

Zhang, Y. B., Tan, Y. W., Stormer, H. L., and Kim, P., 2005, *Nature* (London), 438, 201.

12 Nanostructures— Strategic and Tactical Issues

12.1 THINKING ABOUT STRATEGY

In Chapter 1, the process of materials characterization was explored as an example of strategic and tactical thinking. The following five chapters covered, in broad outlines, the physical principles, technical implementation, and operational modes of the most widely used techniques for nanoscale characterization. The five chapters subsequent to those were framed as illustrations of the general applications of the techniques and methods to various classes of nanostructure. This chapter returns to the strategic and tactical issues that have not been considered elsewhere, and which are relevant to the characterization process.

There can be various entry points to strategic thinking about the characterization process. In most cases, the starting point is that of the structure (e.g., nanotube, quantum dot, etc.) with progress to questions about what information is required. This leads in turn to choice(s) of technique(s), and so on. In relatively rare cases, there is complete freedom of choice of techniques. More often, the choice is constrained by what is available, or by what is affordable.

Sometimes the point of departure is a particular technique (possibly the one that is available and affordable, see above). The thinking is then focussed on the breadth and depth of applications to nanostructural characterization available with a particular technique.

12.2 THINKING ABOUT TACTICS

Tactical thinking about materials characterization is usually much more focussed and specific, but, in general, the thinking can be concerned with a broad range of issues. It encompasses such issues as specimen preparation and mounting, instrument calibration, data acquisition and reduction, recognition of artefacts, and interpretation of 'good' data ('good' data implies that they have relevance to the question(s) being asked and are obtained by an appropriate method by a skilled operator in accord with best practice). The diversity arises from the great variety of specimens and the number of questions that can be asked. Thus, more often than not, the actions required during a first iteration of tactical decisions for nanoscale characterization will involve yet another learning experience. Very rarely is the first image or spectrum the one that gets published or incorporated in a thesis or report. It may even be that the initial

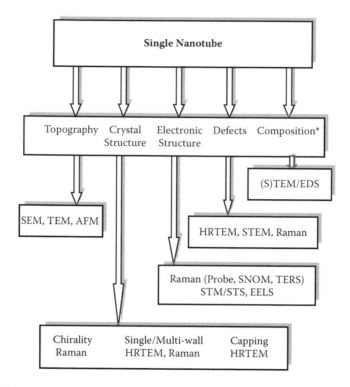

FIGURE 12.1   Overview of the strategic issues for the characterization of a single nanotube. The term 'capping' refers to the termination of the tube. *Compositional information is irrelevant for CNT, aside from the presence of dopants in which case EELS and Raman could provide useful information, but is a significant issue for other tube structures (e.g., BN tubes).

characterization shows that any further such work should be abandoned in favour of returning to the stage of specimen synthesis.

This chapter will confine itself to dealing with the general strategic issues that are most likely to be of concern during a process of nanoscale characterization.

12.3   STRATEGIC ISSUES

The general strategic issues for single nanostructures, such as a nanotube, a quantum dot, or a flake of graphene, can most conveniently be set out in the style of flow-chart diagrams, as shown in Figures 12.1 through Figure 12.3.

12.3.1   OTHER NANOSTRUCTURES

There are, of course, other types of nanostructure, such as the 0-D fullerene cage molecules and 1-D nanowires. For the latter, a strategy for characterization similar to that for a tube structure can be adopted. Topographic information about single wires can be acquired using electron-optical and probe microscopies, crystal structure and structural defects by high resolution transmission electron microscopy (HRTEM),

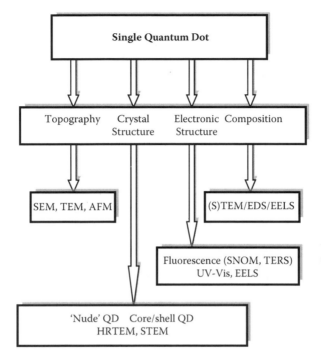

FIGURE 12.2 Overview of the strategic issues for the characterization of a single quantum dot.

composition by energy dispersive spectroscopy (EDS) and electron energy loss spectroscopy (EELS) attached to (S/HR)TEM instruments, while some electronic information can in favourable cases be obtained by EELS.

Characterization of single cage molecules is rather more difficult and is rarely attempted. A notable exception to the rule is the HRTEM imaging of peapod structures, where fullerenes are trapped within the core of a nanotube (see Chapter 9). However, characterization of ensembles of such structures can readily be carried out, as outlined below.

12.3.2 CHARACTERIZATION OF NANOSTRUCTURE ENSEMBLES

Ensembles of nanostructure can take two main forms. In one form, the structures are nucleated and grown on a substrate (e.g., nanotubes and graphene), and the result can be considered to be a nanostructured 2-D specimen of micro, or macro, lateral extent. In the other, a powder specimen can be dispersed on a flat substrate. In both, the traditional surface-analytical techniques can be deployed to good effect. As well, scanning electron microscopy (SEM) and the probe-based imaging techniques, complemented by structural analysis by grazing incidence X-ray diffraction (GIXRD) or Raman can be used.

In other cases, the ensemble can be a bulk powder specimen, again of micro, or macro, dimensions. Many of the well-established particle analysis techniques,

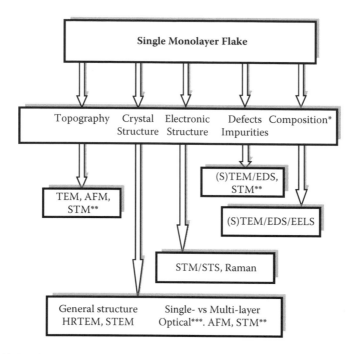

FIGURE 12.3 Overview of the strategic issues for the characterization of a single mono-layer flake. *Composition is irrelevant for carbon-based structures (e.g., graphene and fuller-ites), but important for diatomic flakes (e.g., BN and dichalcogenides). However, the effects of doping carbon-based structures can be inferred from Raman spectroscopy. **Carbon- and chalcogen-based nanostructures are relatively inert, and STM will provide useful results under air-ambient conditions. ***It has been shown that optical microscopy, and Raman analysis, in the reflection mode, can distinguish between single and multilayer flakes. (See recent review, Allen et al., 2010.)

described in Chapter 6, will then be appropriate. Conventional powder XRD can also provide structural information (see Chapter 8).

12.4 PREPARATION OF SPECIMENS FOR CHARACTERIZATION OF NANOSTRUCTURES

In a very general sense, the preparation of nanostructural specimens is remarkably simple. In particular, any 'free' structure that is nanoscale in 1, 2, or 3 dimensions is, by definition, electron-optically transparent in its as-received condition. Likewise, in the cases of SEM or scanning probe microscopic (SPM) analysis, the structures can simply be deposited onto a suitable solid substrate. Raman or fluorescence spec-troscopic analysis in the reflection mode requires a similar deposition of a suitable dilution of the structures on a substrate, or in a suspension. The substrate must be transparent, and preferably spectroscopically inert, if the spectral information is to be collected in the transmission mode.

There is one situation in which a 'traditional' method of specimen preparation must be considered. When nanostructures are nucleated and grown on a substrate, and when either the nucleation and growth mechanisms, or the lateral order or vertical alignment are of interest, then a cross-sectional specimen must be prepared. If the substrate is crystalline, for example a Si wafer, then a cross-section can be prepared by cleavage, and will then be suitable for SEM inspection. If TEM analysis is required, then focussed ion beam (FIB) is a convenient method for preparing an electron-transparent specimen in cross-section. The FIB technique is described in the appendix to this chapter.

The various combinations of structure, technique, and substrate are summarized in Tables 12.1 and 12.2, for single 'free', or for ensembles, of nanoscale objects, respectively.

TABLE 12.1
Single 'Free' Nanoscale Objects

Object	Attribute	Technique(s)	Specimen Preparation
Nanotube	Shape	TEM	Dispersed onto holey carbon film
Nanowire		SEM, AFM, STM	Dispersed onto flat substrate (e.g., Si, mica)
	Crystal structure	(S/HR)TEM	Dispersed onto holey carbon film
		Raman probe	Dispersed onto suitable substrate (e.g., Si, mica), or in a suspension
	Composition	(S/HR)TEM	Dispersed onto holey carbon film
	Dopants	EDS or EELS	
	Electronic structure	(S/HR)TEM, EELS	Dispersed onto holey carbon film
		Raman probe, STM	Dispersed onto flat substrate (e.g., n-Si)
Quantum	Shape	HRTEM	Dispersed onto holey carbon film
Dot		SEM, AFM	Dispersed onto flat substrate (e.g., Si, mica)
	Crystal structure	(S/HR)TEM	Dispersed onto holey carbon film
	Composition	(S/HR)TEM	Dispersed onto holey carbon film
	Dopants	EDS or EELS	
	Electronic structure	SNOM/TERS fluorescence	Dispersed on transparent substrate (transmission)
			Dispersed on opaque substrate (backscatter)
Monolayer	Shape	AFM, STM	Dispersed onto flat substrate (e.g., Si, glass slide, mica)
Flake	Crystal structure	HRTEM	Dispersed onto holey carbon film
		Raman probe	Dispersed onto suitable substrate (e.g., Si, mica), or in a suspension
	Composition	(S/HR)TEM	Dispersed onto holey carbon film
	Dopants	EDS or EELS	
	Electronic structure	(S/HR)TEM/EELS	Dispersed onto holey carbon film
		STM	Dispersed onto flat substrate (e.g., n-Si, HOPG)

TABLE 12.2

Ensembles of 'Free' Nanoscale Objects

Object	Attribute	Technique	Specimen Preparation
Nanotubes	Shape		
	Not oriented	SEM	On growth substrate
	Oriented	SEM	Cross-section on growth substrate
	Crystal structure	Raman	On growth substrate, or dispersed on substrate
		(GI)XRD	Powder specimen, or dispersed on substrate
	Composition	XRF	Bulk powder specimen
		XPS	Powder on conductive foil
	Electronic structure	XPS	Powder on conductive foil
	Surface functionality	XPS	
	Chirality	Raman	Powder on flat substrate
Fullerene Fullerite Films	Shape	AFM	Dispersed powder on flat substrate
			Film on substrate
	Crystal structure	Fullerene: XRD	Bulk powder, or powder on substrate
		Fullerite: Raman, GIXRD	Film on substrate
	Fullerene size distribution	DCS	As for particle size analysis
	Surface functionality	XPS	Fullerene: Powder on conductive foil
			Fullerite: Film on substrate
Monolayer Flakes	Shape	AFM, STM	Dispersed flakes on substrate
	Crystal structure	STM	Dispersed flakes on flat substrate
		GIXRD, Raman	Bulk flakes on flat substrate
	Single/multi-layer	AFM	Dispersed flakes on flat substrate (height)
		Raman	Dispersed flakes on flat substrate (spectrum)
	Composition	XPS	Flakes on conductive foil
	Electronic structure	XPS	Flakes on conductive foil
Quantum Dots	Shape	SEM	As-grown on substrate, or dispersed on substrate
		AFM	As-grown on substrate, or dispersed on substrate
	Crystal structure	GIXRD	As-grown on substrate, or dispersed on substrate
		XRD	Bulk powder
	Composition	XRF	Bulk powder
		XPS	Powder on conductive foil
	Electronic structure	Fluorescence spectroscopy	Bulk powder, or as-grown on substrate, or dispersed on substrate
	Surface functionality	XPS	Powder on conductive foil

12.5 ENSEMBLE AVERAGES: LIMITATIONS

The limitations of methods for particle sizing have been discussed in Chapter 6; in general, the outcome is the average size of an equivalent sphere. Nanostructure ensembles are rarely, if ever, monodisperse. Quantum dots can be synthesised under carefully controlled conditions to be nearly monodisperse, or they can be separated into size fractions by high performance liquid chromatography (HPLC). C_{60} ensembles generally have a significant admixture of the C_{70} size fraction and lesser contributions from other cage structures. Particle sizing based on DLS, or preferably DCS, can resolve multi-disperse ensembles. Given appropriate standards, the results can be quantified.

An endemic problem is the presence of carbonaceous or other types of 'junk'. Some synthesis routes, in combination with purification procedures can eliminate most of the unwanted by-products, but total purity is never guaranteed.

In cases where there is a near-continuous distribution of attributes for an ensemble, such as size, chirality, composition, surface chemistry, and so forth, it must be kept in mind that the result of characterization will be a weighted average over the information volume of the technique. Likewise, when a nanostructured specimen is investigated by a technique with an information volume of greater dimension than that of the structural unit(s), the result will be a weighted average of whatever is within the information volume. Nevertheless, sometimes it is possible to deconvolute the results, such as in the case of spectral information, and thus to gain useful information about the nature and extent of dispersion of the sizes and other attributes of an ensemble of nanostructural units.

12.6 'SOFT' MATERIALS—SPECIMEN PREPARATION

Studies of nanostructured soft materials, such as polymers (e.g., phase-separated diblock copolymers) and biomaterials (e.g., tendons, molecular motors, actin filaments, etc.,) as well as many naturally occurring nanostructures, have become fields in their own right. There is an increasing trend in the case of 'soft' materials towards deployment of the same techniques and methods, such as SEM, TEM, SPM, and Raman, which have been used successfully in the wider field of nano-science and -technology.

In some cases, specimen preparation can be relatively straightforward. For instance, AFM imaging and F-d analysis are generally undertaken on as-received specimens, excised from a bulk structure. Likewise, SEM imaging and Raman spectroscopy require minimal specimen preparation.

TEM analysis, on the other hand, will require special preparation techniques, in order to produce electron-transparent foils. Slicing into thin sections with a (cryo) microtome instrument, although time-consuming and onerous, remains the most widely used procedure. A brief overview of the method is given in Appendix B.

12.7 CLEANLINESS

It is impossible to overstate the case for cleanliness in materials characterization, in general, and for nanostructures, in particular. Aside from the presence of junk arising from the synthesis routes, other forms of adventitious contamination must be avoided as much as possible, especially if surface analytical or scanning probe techniques are being deployed. Nanostructures are essentially composed primarily of surfaces (e.g., SWCNTs and monolayer flakes are 100% surface). Thus, contamination, if present, will in extreme cases constitute most of the information volume.

Nanostructures generally arrive from a supplier as either a dry powder or in solution. In the former case, agglomeration will be a serious problem, unless the powder is subsequently dispersed before characterization in a non-reactive, non-polar fluid (e.g., alcohol). In the latter case, a surfactant may have been added to a buffered solution, in order to prevent agglomeration, and in some instances, an antibiotic or antifungal agent will have been included, in order to suppress bacterial or fungal growth. The additives constitute contaminants from the point of view of characterization. Under electron irradiation, the organics will turn into carbonaceous contamination, while in surface analysis, the surface chemistry will be masked.

12.8 USER-FRIENDLINESS

At the stage of basic research, or in cases where characterization is a one-off activity, user-friendliness is unlikely to be a significant issue. In both instances, gaining access to the most effective technique and method, in combination with relevant expertise, irrespective of user-friendliness and cost, will be sought.

The numbers along the 'availability' axis in Figure 12.4 represent rough estimates of the number of instruments of a particular type installed worldwide. For

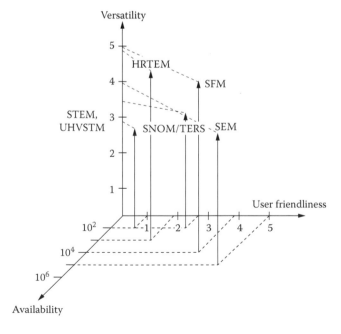

FIGURE 12.4 Schematic representation of the relationships between user-friendliness, availability, and versatility, for some of the most widely used techniques for characterization of nanostructures.

instance, the world-wide inventory of SEM instruments of all types and capabilities is likely to exceed 10^5. However, a significant fraction of that inventory may not be suitable for nanoscale imaging. On the other hand, the inventory of UHV-STM facilities is likely to be of the order of 10^2. User-friendliness is defined by a scale from 0 to 5, with 5 being the most user-friendly (that is, a training session of a couple of hours should suffice to get a novice to the point of being able to obtain usable images, and in which the interpretation of the images would be relatively intuitive). On the other hand, UHV-STM and HRTEM techniques involve sophisticated instruments requiring a lengthy period of training, and the interpretation of images and spectra is non-trivial. The index 'versatility' is related to the breadth of applications for the most widely studied nanostructures, and to the range of attributes that can be interrogated with a single technique. For instance, HRTEM scores highly because it can be applied to all known nanostructures, and because, with appropriate attachments, it can generate information about shape, crystal structure (including defect structure), elemental composition (by EDS), and electronic structure (by EELS).

12.9 COST-EFFECTIVENESS

The determination of the difference in cost-effectiveness between investing in, using an existing, or commissioning the use of, a particular technique, will depend on a complicated evaluation of a number of inter-related factors.

Arguably, the most important consideration is that of the depth and breadth, and the duration of nanoscale characterization activities. A couple of hypothetical, possibly extreme, examples will suffice to illustrate the point.

Example 1. A small start-up company is developing a novel synthesis process for single wall nanotubes (SWCNTs); the programme is expected to have a duration of two years. The requirement is for structural characterization of single tubes. In this example, there is a very strong case for contracting out HRTEM analysis, even at the cost of £1000/day for access to instrument and expertise. The alternative is to invest £300,000+ in an instrument, plus infrastructure. Additionally, the expertise and a maintenance contract would add around £100,000 per year. The delays in ordering and installing the facility, employing the expertise, and establishing procedural protocols could impose yet more disincentives.

In due course, a large multinational company takes over the process, which now promises to be commercially viable, for producing SWCNTs in bulk quantities. The said company needs to carry out online quality control of the SWCNT product. The decision is made to invest in a fully automated flow-through Raman spectrometer facility, at a capital cost of ca. £400,000. The alternative would be to contract out batches of samples, at £250 per sample. Allowing for two batches per day the simple economics may not be persuasive. However, the turn-around time, and loss of control, may swing the argument in favour of having an in-house capability.

TABLE 12.3

Cost-Effectiveness of Techniques for Nanoscale Characterization

Technique	Capital Cost[a] (£K)	Versatility[b]	Recurrent Requirements[c]
SEM	50 (entry level) 250+ (top of range)	Excellent	Low level of expertise and low maintenance
HRTEM	300 (entry level) 3000 (top of range)	Good	High level of expertise Needs maintenance contract
Raman	50 (entry level) 300+ (top of range)	Excellent	Low/medium level expertise
SPM (air)	50 (entry level) 250+ (top of range)	Excellent	Low/medium level expertise Medium level expertise
UHVSPM	500+	Modest	High level of expertise and high maintenance
SNOM/TERS	150+	Modest	Medium level of expertise

[a] The capital costs of techniques are estimates at current list prices at present exchange rates (mid-2011). The estimates are strongly dependent on the choice of attachments to the base models. The estimates do not include infrastructure requirements, such as special accommodations (e.g., vibration isolated floor space), air conditioning, chilled water, and so forth.

[b] The versatility index refers to the breadth of applications of a particular technique to the range of the most topical nanostructures, and to the ability to probe more than one variable.

[c] Recurrent costs relate principally to the salary/ies of technical expertise required to maintain and develop an instrumental facility, and/or the expertise required to operate the facility to best effect and to interpret the output. Other relatively minor costs are concerned with operational consumables, delivery of infrastructure consumables, routine maintenance, and repairs.

Example 2. For a tertiary teaching institution, or a research institute, the arguments are far more complex. A central instrumental facility will then service a great range of needs, some of which cannot readily be quantified, such as teaching support, basic research, development of skills and expertise, and prestige. In general, universities can find the funding from time to time for the capital investment, but more often than not be unwilling or unable to underwrite the recurrent cost of maintaining its effective operation.

Some indices that will affect thinking about the cost-effectiveness of the most widely used techniques for nanoscale characterization are summarised in Table 12.3. Not surprisingly, techniques, such as SEM, with low capital cost, high versatility, low demands on expertise, and minimal recurrent costs are also the most widely available.

ACKNOWLEDGEMENTS

We wish to record our appreciation to Dr. C. Johnston for many useful conversations and helpful comments.

APPENDIX A: PREPARATION OF CROSS-SECTIONAL SPECIMENS BY THE FOCUSSED ION BEAM (FIB) METHOD

GENERAL DESCRIPTION

Focussed ion beam (FIB) techniques and methods bring together the capabilities on one instrumental platform for imaging, manipulating, modifying, and analysing structures of nanoscale dimensions. In the context of specimen preparation, FIB is now used to prepare cross-sectional specimens for high-spatial resolution analysis by electron-optical and scanning probe techniques.

The main elements of an FIB instrument include: a liquid metal fine-focus ion gun (LMIS) for nanoscale erosion; an SEM column for imaging the erosion process in real time; an eucentric sample stage capable of translation, rotation, and two-axis tilt; and software-programmable control of electronics for spatial control of the specimen location and orientation, as well as control of the location and current of the ion beam. A distinctive feature of FIB systems is the facility for the feeding in of metal-organic precursor gases. This allows physical vapour deposition (PVD) of spatially resolved sacrificial, or protective, layers.

FIB-assisted depostion takes place when a precursor gas, such as tungsten hexacarbonyl ($W(CO)_6$) is injected into the working chamber. When scanning an area of interest with the ion beam, the chemisorbed precursor gas decomposes into volatile and non-volatile components. The non-volatile component, such as W, remains as a deposited layer on the surface. Other materials, such as Pt, Co, C, Au, and so forth, can also be deposited as protective/sacrificial layers.

The erosion and imaging processes are shown schematically in Figure 12.5, where the focal points of the ion and electron beams coincide on the specimen. While the ion beam is responsible for the erosive sputtering, both beams give rise to emission of secondary electrons, and can generate a high-resolution image.

TYPICAL INSTRUMENT

Figure 12.6 shows an example of a current generation dual beam FIB instrument. The specifications of the instrument are summarised in Table 12.4.

SPECIMEN PREPARATION BY FIB METHODS

Advantages

The combination of ion beam milling at high spatial resolution, 10 nm, and SEM imaging, offers much improved resolution when selecting a region of the sample, such as a grain boundary, in comparison with other specimen preparation techniques, (e.g., thin sections can be prepared to an accuracy of better than 20nm).

FIB is virtually independent of the nature of specimen material, although the sputtering rate will depend on the type of specimen, and on its composition. In particular, grains in multi-phase composites may sputter at different rates. Good insulators can pose a problem unless charge neutralisation is being used.

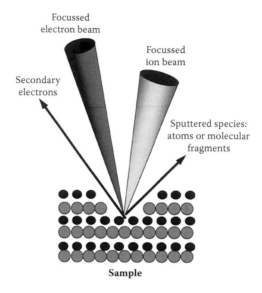

FIGURE 12.5 (See color insert.) Schematic depiction of the dual beam arrangement. Incident ions sputter away atoms or molecular fragments from the point of focus. Secondary electrons are ejected from the respective focal points of the electron and ion beams.

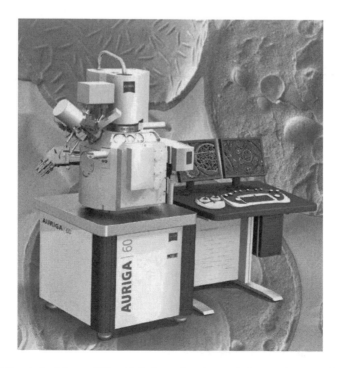

FIGURE 12.6 A dual-beam FIB with independent ion and electron beam optical columns. (Zeiss Auriga 60, http://download.Zeiss.de/Auriga60/Auriga_60Flyers.pdf.)

TABLE 12.4

General Specifications for the Zeiss Auriga FIB Facility

	SEM	FIB
Lateral Resolution (SEI)	1.0 nm @ 15 kV	< 2.5 nm @ 30 kV
	1.9 nm @ 1 kV	—
Magnification	×12 to ×10^6	×300 to ×500,000
Probe Current	4 pA–20 nA	1 pA–50 nA
Acceleration Voltage (kV)	0.1–30	< 1–30
Source	Schottky field	Ga liquid metal

System	
Gas Injection	(a) Five precursors (Pt, W, C, F, etc.)
(Options)	(b) Four precursors, plus charge compensation
	(c) Single precursor
	(d) Automated and retractable injection, charge compensation, and sample cleaning
Sample Stage	6-axis motorised eucentric
	X-Y range: 152 mm
	Z range: 13 mm
	Tilt: −15–70°
	Rotation: 360° repetitive
Detectors	SEM: In-lens annular and chamber SE
	FIB: Various options

Source: Adapted from the Zeiss Web site.

When compared with more conventional methods, the FIB technique is faster and more reliable. Depending on the specimen material and its structure, sample preparation times can be as short as 20 min, and are rarely more than a few hours in duration.

Disadvantages

As always, there are some disadvantages. Irradiation with metal ions at kinetic energies in the keV range will modify the surface layers in several significant ways. First and foremost is the implantation of Ga atoms; in extreme cases, the implantation can lead to nucleation and growth of embedded droplets. Secondly, the dose rates per unit area will result in energetic collisions between the incident species and every atom in the first few layers of the surface. The result is the formation of a highly disordered surface layer. Sometimes, partial recrystallisation can be achieved by heat treatment. Finally, due to the great variation in sputter yields of different surface species, after energetic ion bombardment, the composition in the first few monolayers will be significantly different from that of the bulk. In particular, oxides are usually reduced as a result of ion bombardment, resulting in a surface layer that is both subvalent and non-stoichiometric. If subsequent EELS analysis were to be carried out, then the effects of bombardment on the near-surface layers would have to be taken into account.

Details of FIB Methods

In the context of nanostructures, FIB methods are particularly important for electron-optical or scanning probe analyses, in situations where the nanostructures are embedded in, or attached to, larger scale materials or devices. The role of the FIB is to reveal and isolate the structures, or the interface, for study by SEM, TEM, or SPM.

Several sample preparation methods have been developed, the so-called H-bar technique being among the earliest. It was modified subsequently and the variations on it became known as the *ex situ* lift-out (EXLO) and the *in situ* lift-out (INLO) methods. The technique was developed initially for the study by TEM of semiconductor devices, and was applied subsequently to other materials and structures. After an initial mechanical polish, FIB is used to cut trenches on either side of the region of interest, as shown in Figure 12.7(a) and (b). The section for TEM analysis is the 'bar' in the H-shaped structure.

The INLO version of the technique is described schematically in Figure 12.8. In (a) a selected region in the sample surface has been machined into a wedge shape by FIB, and a micromanipulator probe arm is about to lift the wedge from the surface. The region is observed via a SEM. Once the end of the probe is attached to the wedge with metal straps deposited *in vacuo*, the wedge can be transferred to a carrier, in this case a TEM grid, as in (b). Further metal straps are then deposited to fasten the wedge to the grid, following which, the probe-to-sample straps are removed by ion beam milling, resulting in the situation shown in (c). The grid is then tilted, as in (d), so that the wedge can be thinned by FIB until it is electron-transparent, leading to the H-configuration in (e). Finally, the grid is rotated back to its original position (i.e., as in (b) and (c)), and transferred into the TEM for analysis. In the EXLO version, the sample is removed from the FIB chamber after the wedge has been formed.

Subsequent development of the H-bar technique has tended to dispense with the initial polishing stage. Generally, the procedure begins with the deposition of a protective layer of W or Pt onto the area to be thinned. The ion beam then removes material from either side of the W or Pt patch.

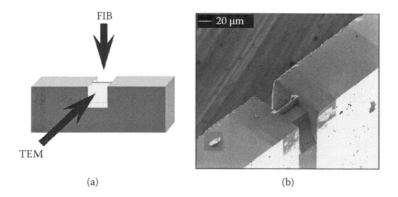

(a)   (b)

FIGURE 12.7   (a) The schematic shows the direction of the ion beam for creating two trenches. (b) A SEM image of the actual structure. (From Mayer et al, 2007, *MRS Bull,* with permission from the Cambridge University Press.)

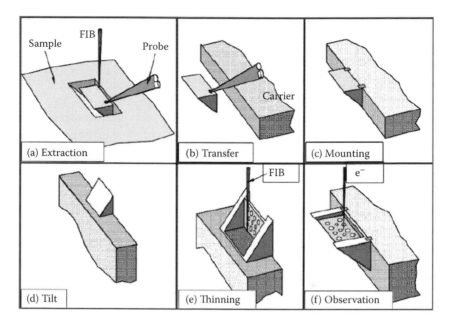

FIGURE 12.8 **(See color insert.)** Schematic illustrations of the stages in the extraction, transfer, mounting, and final preparation by the wedge method of a sample for TEM analysis, by the INLO technique. (From Mayer et al., 2007, *MRS Bull,* from the Cambridge University Press, with permission.)

Additional detail can be found in the recent literature (e.g., Li et al., 2006; Volkert and Minor, 2007).

APPENDIX B: (CRYO)MICROTOMY AND OTHER METHODS FOR SPECIMEN PREPARATION OF SOFT MATERIALS (E.G., POLYMERS AND BIOMATERIALS)

GENERAL OVERVIEW

Some soft materials can be studied by SEM in their as-received condition. However, more commonly, it is necessary to deposit a thin conductive coating, so as to avoid charging effects, in order to obtain the best results. The water content in as-received biomaterials presents a particular problem. When a sample of that type experiences a sudden exposure to vacuum (i.e., in the sample position in a SEM or TEM), the rapid loss of moisture can cause major changes to shape and structure, and thus invalidate the analysis. Such specimens usually need to be fixed and dehydrated *ex situ* before SEM analysis can be attempted. Furthermore, the sudden increase in water vapour pressure may lead to instrumental deterioration or failure.

AFM analysis under air-ambient conditions, or in a fluid cell, is much more forgiving of the state of a biomaterial specimen. In many cases it is possible to image 'living' tissue at molecular resolution. The availability of the AFM tapping mode allows sensing of out-of-plane force components, while the in-plane shear forces are effectively turned off, thus eliminating the major cause of tip-induced distortion and damage of extremely soft and/or fragile materials, such as a cellular plasma membrane.

One of the most widely used techniques for the preparation of samples used in optical and electron microscopy analysis is that of microtomy, which involves the slicing of the specimen into sections thin enough to be transparent to incident photons or electrons. Microtomy, in the ultra- or cryo-version, is the major step in the preparation procedure, but is generally followed by other steps, such as fixation, dehydration, and chemical treatment to improve contrast. The venerable classic monograph by Hall (1953) remains one of the best sources for information about the procedures.

Each slice, of thickness in the 100 nm range, from a 3-D object, will represent a 2-D section of a larger structure. The analysis of successive slices can therefore be viewed as a method of profiling, thus revealing structure and composition throughout the 3-D volume. Recent developments in computer image processing allow the reconstruction of the specimen by combining the sequence of 2-D images into a 3-D image.

TYPICAL INSTRUMENT

A typical instrument is shown in Figure 12.9. Several different types of microtome instrument have been developed, including the so-called sled and rotary types. The common critical element is a sharp knife made of either glass or diamond. A programmable motorised stage, with a spatial resolution in the low nm range, moves the sample against the edge of the cutting knife. The process can be viewed through a binocular microscope. Successive thin slices of specimen material are carefully lifted away from the knife and deposited on a suitable holding substrate. The serial sections are then transferred to a TEM grid for imaging and analysis (see Figure 12.10).

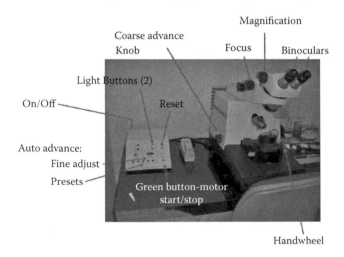

FIGURE 12.9 (**See color insert.**) A typical microtome. The knife and sample holder are viewed through a binocular microscope. (From http://sbs.wsu.edu/fmic/protocol/microtome.ppt.)

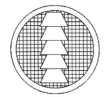

One large section per grid   A ribbon of serial sections   As many as possible

FIGURE 12.10 Three possible options for mounting microtomed slices onto a TEM support grid. (From http://sbs.wsu.edu/fmic/protocol/microtome.ppt.)

Some 'soft' materials are affected adversely by the slicing procedure. The resultant damage can be reduced by cooling the specimen and knife to a temperature below the glass transition temperature of the sample material, which transforms the normally soft and plastic sample to a hard and brittle state. This variation on the mictrotome theme is known as cryo-microtome or cryo-ultramicrotome.

REFERENCES

Allen, M. J., Tung, V. C., and Kaner, R. B., 2010, *Chem. Rev.*, 110, 132.
Hall, C. E., 1953, *Introduction to Electron Microscopy*, New York: McGraw-Hill, http://sbs.wsu.edu/fmic/protocol/microtome.ppt.
Mayer, J., Giannuzzi, L. A., Kamino, T., and Michael. J., 2007, *MRS Bull.*, 32, 400.
Li, J. Malis, T. and Dionne, S., *Materials Charact.*, 2006, 64.
Volkert, C. A. and Minor, A. M., 2007, *MRS Bull.*, 32, 389.

Index

A

Absorption and photoluminescence
spectroscopies, quantum dots,
"blinking," 208–209
spectra, 206–208
ADF (Annular dark field imaging), STEM, 29
Adsorption and desorption, 154–161
rate constants, 154–156
Adsorption isotherms, 157–159
Asorption, sequential fractional coverage,
154–155
AFM (atomic force microscopy), biomolecular
binding and unfolding, 107–110
CNTs, 238–239
colloidal probe analysis, 107
deformation of collagen fibril,
nanoindentation, 107–108
nanoindentation, 105–107
nanosheets, topography, 277, 281
nanowire FET test device, 266, 268
single atom chemical identification, 128, 130
topographic imaging, quantum dots, 201–202
Arc-discharge growth, CNTs, 218
Aspect ratio, SFM probe, 87
Auto-correlation, nanoparticle size distribution,
DLS, 139–143

B

Band structure, SWCNTs, 223–224
BET expressions, 157–160
BET method, surface area measurement, 154–161
BET particle size analysis, equivalent spheres,
161
Biomolecular applications, fluorescence
spectroscopy, 69–70
Biomolecular binding and unfolding, F-d
analysis, 107–110
"Blinking," photoluminescence, quantum dots,
208–209
BNNTs, EELSspectra, 233
HRTEM images, 232–234
SAD patterns, 232
SEM image, 232
Boron nitride, nanotube (BNNTs), 223–224
nanosheets, compositional information,
STEM, 277, 280
structural characterization, 277, 280–281
synthesis, liquid phase exfoliation, 275

C

Cage structures, fullerenes, 173–174
Calibration, F-d analysis, 104–105
methods for normal force constant, SFM, 131
SFM probe, 88
Capacitance mapping, SCM, 110–111
Carbon allotropes, Raman spectra, 71–74
Carbon, bonding considerations, 271
SRXPS spectra, grapheme, 285–286
Catalyst-promoted nucleation, synthesis of
graphene, 276–277
Catalysts, use in synthesis of CNTs, 218–219
Characterization limitations, nanostructure
ensembles, 295
Characterization parameters, nanoparticulate
materials, 135–136
Characterization requirements, CNTs, 225
Characterization strategy, nanostructures,
290–292
Chemical vapour deposition (CVD), synthesis of
CNTs, 218–219
Chemisorption, definition, 154
Cleanliness requirement, nanostructure
characterization, 295–296
CNTs (carbon nanotubes), AFM, 234, 238–239
"armchair" structure, 216, 223
BN-coated, EELS characterization, 245–247
characteristics, comparison with other
materials, 224
characterization requirements, 225
chirality, 215–216
"chiral" structures, 216, 223, 238
description, 215–218
electrical conduction, 223–224
electronic structure, 219–223
field emitters, 227–229
formation from graphene, 271
purification of synthesised product, 219
quality control procedures, 226
Raman spectroscopy, 70–74, 238–243
STM, 234, 237
structural characterization, TEM, 226, 228,
230–231